World Cities and Climate Change

Issues in Society
Series editor: Tim May

Published titles

Zygmunt Bauman: *Work, Consumerism and the New Poor, 2nd edition*
David Byrne: *Social Exclusion, 2nd edition*
Graham Crow: *Social Solidarities*
Mitchell Dean: *Governing Societies*
Gerard Delanty: *Citizenship in a Global Age*
Steve Fuller: *The Governance of Science*
Mike Hodson and Simon Marvin: *World Cities and Climate Change: Producing Urban Ecological Security*
David Knights and Darren McCabe: *Organization and Innovation: Guru Schemes and American Dreams*
Nick Lee: *Childhood and Society*
David Lyon: *Surveillance Society*
Linda McKie: *Families, Violence and Social Change*
Pam Odih: *Work, Time and Identity in a Cultural Context*
Graham Scambler: *Health and Social Change*
Graham Scambler: *Sport and Society: History, Power and Culture*
Piet Strydom: *Risk, Environment and Society*

World Cities and Climate Change

Producing Urban Ecological Security

MIKE HODSON AND SIMON MARVIN

McGraw Hill
Open University Press

Open University Press
McGraw-Hill Education
McGraw-Hill House
Shoppenhangers Road
Maidenhead
Berkshire
England
SL6 2QL

email: enquiries@openup.co.uk
world wide web: www.openup.co.uk

and Two Penn Plaza, New York, NY 10121-2289, USA

First published 2010

A catalogue record of this book is available from the British Library

ISBN-13: 978-0-33-523730-2 (pb) 978-0-33-523729-6 (hb)
ISBN-10: 0-33-523730-4 (pb) 0-33-523729-0 (hb)

Library of Congress Cataloging-in-Publication Data
CIP data applied for

Typeset by Aptara® Inc., India
Printed in Great Britain by CPI Antony Rowe, Chippenham, Wiltshire

The McGraw·Hill Companies

Contents

Figures

Tables

Acknowledgements

This book has been developed over a number of years. Many people and organizations have provided support for this project and assisted us financially, logistically, intellectually and emotionally. We would like to express our thanks and gratitude for that support.

We would like to thank those organizations that have financially supported the research, which has been fundamental to the production of the book. In particular we would like to acknowledge the following: the European Commission for its funding through a number of Framework programmes; in the UK we have received funding from the Engineering and Physical Sciences Research Council (EPSRC), the Economic and Social Research Council (ESRC), the Northern Way and the Environment Agency; the University of Copenhagen and the MISTRA foundation in Sweden provided support for the internationally comparative work; and work with the global engineering company Arup provided a commercial perspective on ecological urbanism.

Our key argument is the product of a process of testing propositional knowledge that has been refined through numerous presentations, different styles of writing, dialogue and discussion with researchers and policymakers. The initial ideas developed in this book were presented at an Open University workshop on the Politics of Infrastructure, in June 2006, organized by Colin McFarlane, when an early draft of our ideas was produced. Jonathon Rutherford and Olivier Coutard encouraged us to develop and publish our argument. A further development of the thesis was presented at the Reproducing City-Regions international workshop, organized by SURF, in Manchester in July 2007. We are grateful for the constructive comments of participants, particularly those from Aidan While, Harriett Bulkeley and

Bob Jessop. We are appreciative of the constructive responses to our arguments from those who have funded our research and from policymakers in Manchester, Leeds, London, Melbourne and Tokyo. The development of our ideas have also benefited from our discussions in Paris, Berlin, Graz and Manchester with colleagues in the Urban Infralab, in particular Olivier Coutard, Jonathon Rutherford, Tim Moss, Matthias Naumann, Harald Rohracher and Simon Guy. We have presented versions of this argument to academics in Manchester, London, Lancaster, Zurich, Melbourne, Tokyo and Boston. We would like to thank participants in these discussions for their many critical and constructive comments on earlier versions of our argument. Steve Graham and Maria Kaika gave us very constructive and encouraging feedback.

We appreciate Roger Keil's invitation to provide the inaugural paper in the Urban Worlds section in the *International Journal of Urban and Regional Research* (Hodson and Marvin 2009a) – where the ideas developed in this book were further refined. Other parts of our argument were also published in *International Journal of Urban and Regional Research, Technology Analysis and Strategic Management* and *Research Policy* (Hodson and Marvin 2007; 2009b; 2010).

We have been extremely pleased and privileged to be able to produce the book within the supportive and collaborative context of SURF. We are particularly grateful to our colleagues – Tim May, Beth Perry and Vicky Simpson – not only for this support but also for their friendship. Vivian Liang provided excellent support in sorting out the references, figures and permissions. Tim, both as a friend and colleague and as editor of this series, has contributed through his intellectual interests and constructive support to many of the arguments contained within the book.

Finally, we would like to offer our thanks to our families – Sarah, Daniel and James Hodson and Nicola Turner and Lewis Marvin – for their love and support during the writing of the book.

We would like to acknowledge the following for permission to reproduce figures and tables. Dr Mark Williams of the University of Leicester for Figure I.1, Cover of *Geological Society of America*. The OECD for Figure 2.1, the map of major city climate change studies. Floodsim for Figure 2.2, The effects of climate change are laid bare in an online simulation which shows gamers the impact of flooding on London. The Federal Communications Commission for Figure 2.3, Interconnected infrastructure networks. Defra for Figure 2.4, Urban carbon emissions in the UK. Arup for Figures 3.1 and 3.2, artistic impressions of Treasure Island and Dongtan. DCLG for Figure 3.3, Thames Gateway 'eco-region' near London. California Hydrogen Highway for Figure 3.4, California Hydrogen Highway. Fuel Cell Bus Club for Figure 3.5, New hydrogen transport networks. World Wildlife Fund for Figure 3.6, Conceptual framework and key drivers for the low carbon city. City of Portland, Oregon for Figure 3.7, Peak Oil City Portland. Transition Towns for Figure 3.8, Transition Towns. Edward Cohen-Rosenthal and Thomas N. McGalliard for Figure 3.9, Symbiotic relationships between the core partners of Kalundborg. Local Power for Figure 3.10, Energy independence for San Francisco. The C40 for Table 4.1, C40 cities, targets and technologies, Table 4.2, C40 cities, and Table 4.3, C40 affiliate cities.

SURF, Manchester, December 2009

Series editor's foreword

The social sciences contribute to a greater understanding of the workings of societies and dynamics of social life. They are often, however, not given due credit for this role and much writing has been devoted to why this should be the case. At the same time we are living in an age in which the role of science in society is being re-evaluated. This has led to both a defence of science as the disinterested pursuit of knowledge and an attack on science as nothing more than an institutionalized assertion of faith with no greater claim to validity than mythology and folklore. These debates tend to generate more heat than light.

In the meantime the social sciences, in order to remain vibrant and relevant, will reflect the changing nature of these public debates. In so doing they provide mirrors upon which we gaze in order to understand not only what we have been and what we are now, but to inform ideas about what we might become. This is not simply about understanding the reasons people give for their actions in terms of the contexts in which they act, as well as analysing the relations of cause and effect in the social, political and economic spheres, but about the hopes, wishes and aspirations that people, in their different cultural ways, hold.

In any society that claims to have democratic aspirations, these hopes and wishes are not for the social scientist to prescribe. For this to happen it would mean that the social sciences were able to predict human actions with certainty. This would require one theory and one method applicable to all times and places. The physical sciences do not live up to such stringent criteria, while the conditions in societies which provided for this outcome would be intolerable. Why? Because a necessary condition of human freedom is the ability to have acted otherwise and to imagine and practice different ways of organizing societies and living together.

It does not follow from the above that social scientists do not have a valued role to play, as is often assumed in ideological attacks upon their place and role within society. After all, in focusing upon what we have been and what we are now, what we might become is inevitably illuminated. Therefore, while it may not be the province of social scientists to predict our futures, they are, given their understandings and equal positions as citizens, entitled to engage in public debates concerning future prospects.

This international series was devised with this general ethos in mind. It seeks to offer students of the social sciences, at all levels, a forum in which ideas are interrogated in terms of their importance for understanding key social issues. This is achieved through a connection between style, structure and content that is found to be both illuminating and challenging in terms of its evaluation of topical social issues, as well as representing an original contribution to the subject under discussion.

Given this underlying philosophy, the series contains books on topics which are driven by substantive interests. This is not simply a reactive endeavour which reflects dominant social and political pre-occupations, but is also proactive in terms of an examination of issues which relate to and inform the dynamics of social life and the structures of society that are often not part of public discourse. What is distinctive about the series is interrogations of the assumed characteristics of contemporary times in relation to their consequences for the organization of society and social life and appropriate modes of study.

Each contribution contains, for the purposes of general orientation, as opposed to rigid structure, three parts. First, an interrogation of the topic which is conducted in a manner that renders explicit core assumptions surrounding the issues and/or an examination of the consequences of historical trends for contemporary social practices. Second, a section which aims to 'bring alive' ideas and practices by considering the ways in which they directly inform the dynamics of social relations. A third section then moves on to make original contributions to the topic. These encompass possible future forms and content, likely directions for the study of the phenomena in question, or an original analysis of the topic itself. Of course, it might be a combination of all three.

Mike Hodson and Simon Marvin have produced a book with this ethos in mind. In recent years urban studies has been a major intellectual arena in which alternatives to the idea of nation-state control have been articulated through debates on globalization and world city formation. The city is a field of encounters and exchanges among individuals and groups who accord and are accorded different levels of significance and recognition within its space. The nature and boundaries of the city are continually being reconfigured as key economic terrains in their own right with consequences for the environment and quality of life. Particular economic conditions are appealing to those who seek to place them on the 'world stage', while these aspirations play into the hands of the idea of 'flows', 'connections' and 'mobilities' in which the assumed benefits of the increasing commodification of things, people and places are a long way from being fairly distributed among their populations.

In the face of these issues in an era of climate change and resource constraint, we need better understandings of contemporary urban transitions and their consequences. Cities can be seen as both victims of, and contributors to, climate change, as well as places from which innovative responses may emerge. Given these conditions and pressures, together with the need for social cohesion within an increasingly unequal world, how are urban elites responding to the challenge? As world cities (defined here in terms of those that function as key nodes in the global economic system) seek to position themselves in an international hierarchy, what is the form and content of their responses?

As the argument in this study unfolds, different scales of governance over territory are at work and world cities can, and often do, act in ways that transcend their national boundaries in the name of seeking economic gain. Varying forms of anticipation to threats, mitigations and adaptations according to particular constructions of the reproduction of economic necessity are apparent in the strategies pursued. Important issues are then raised concerning sufficiency of resources, how the reconfiguration of socio-technical networks is manifest in these cities and how forms of adaptation to perceived and real threats can explain the degrees of variation in autonomy and self-reliance between cities?

Through an examination of the responses of world cities, with particular emphasis upon London, New York, Tokyo, Shanghai, Melbourne and San Francisco, the authors examine how strategies to secure their material advantage are pursued via particular constellations of social interests, resources and knowledges. Flows of energy, water, waste and people through and within cities are part of politicized processes with varying impacts upon populations and the environment. A bounded idea of space, as that which is hermetically sealed off from its surroundings, can enable cities to exercise indifference to the consequences of their actions for those areas and persons that lie outside of particular places. In the face of increasing process of neoliberalization, this contributes to an overall atomization of provision, leaving the possibility of more relational and strategic views that take account of environmental and social concerns, in their wake.

What clearly emerges from this study is a separation between understandings of urban experience, sustainability, governance, planning and policy-related work. Where they should converge and integrate for general benefit, they are diverging according to narrow, institutional logics. However, simple separations between these issues are no longer possible if effective and sustainable solutions are to be found. Despite this evident need, the authors chart a move towards a technocratic view and away from a relational understanding within and between cities at varying scales of action. An urban ecological security viewpoint then emerges as dominant which views urban sustainability as being similar in form to military and economic security.

World cities are politically and economically positioned to protect what are seen as vulnerable infrastructures in the name of reproducing economic success. Issues associated with decreasing natural resources then meet with actions that aim to secure the status quo at global, through to national, regional and city-regional levels. Accompanying these strategies is a clear tendency to emphasize particular solutions according to narrowly prescribed problems. Predominant issues of transferability and marketization mean that context-sensitive sustainable solutions take a back seat to those that can float freely from place within a globalized and commodified space in which the price, but certainly not the value, of everything is known.

Mike Hodson and Simon Marvin take several examples in charting the form of these responses. Eco-cities provide for a utopian vision of possibility against the reality of actually existing infrastructures that allows the free circulation of an idea without content. They also examine hydrogen highways, eco-towns, low and carbon neutral cities and the process of retrofitting, as well as ideas of energy independence. Piloting new eco-building demonstrations, improving the energy efficiency of existing buildings and public engagement processes for demand reduction and savings are part of the ways in which cities seek to address resource constraint and climate change.

The authors do acknowledge that some practices within world cities are potentially innovative responses from which others can learn, but a dominant emphasis upon resilience and security is typical. Cities are seeking to customize infrastructures with resulting degrees of autonomy from central networks, but leave aside questions over a more generalized and joined-up approach that fuse environmental and social sustainability. A selective logic is apparent in which competitors and newer entrants develop intermediary functions on the back of markets and global transfer with resulting consequences for the practicalities of how to make initiatives work in places using more broad coalitions of social interests. The promotion of self-sufficiency in relation to world cities brings together the economic and ecological through such ideas as carbon neutrality, zero waste and energy independence.

What emerges from this book is the pursuit of an exemplary politics fuelled by global economic hierarchies. Particular places make claims to their solutions as models that others should follow through a fusion of political positioning, technological transfer and commodification. A collective myopia over the fundamental issue of how these solutions may be brought into being tends to be sidelined, in favour of the encouragement of the visibility of 'fixity' to demonstrate that technologies, techniques and approaches are somehow replicable to other contexts without taking the importance of place seriously. Once again, we see a de-coupling around bounded space, rather than a more relational viewpoint. These cities are seeking to become more independent by withdrawing from wider contexts – despite those spaces being bound up at various levels with the existence of the city itself. Such extra-scalar configurations are symptomatic of a new era in climate change politics.

The authors thus conclude their investigations on the implications of secure urbanism and resilient infrastructure, with a set of alternative ways forward. We need more nuanced views of how vulnerable populations are affected by current pressures. Aside from the need to think of space in more relational terms, this would also accept that cities are both victims of, and contributors to, climate change. Rendering infrastructures more visible in terms of how they are configured according to different interests are needed in order that alternative possibilities may then arise. This is essential so that their invisibility and overall sustainability is more apparent to the populations that they serve. These should run alongside broader coalitions of people and organizations to think through issues about an urban quality of life and a move away from technological, production-led fixes to cultural and behavioural transformations against the backdrop of appropriate levels of governance arrangements. Overall, the 'what' of responses needs to be countered by 'how' sustainable solutions are made work in different contexts. That means taking existing conditions and the capacities and capabilities to make a difference in places much more seriously than is apparent in current responses.

It is the pursuit of exceptionality, rather than working to produce collective sustainable solutions that tends to rule current responses and this acts as an impediment to the creation of sustainable solutions. The ways forward that the authors suggest requires the avoidance of this exemplary politics of positioning by world cities within a narrowly constituted economic agenda to enable more open and reflexive processes to produce effective responses. These suggestions need to be taken seriously and that means abandoning the narrowness of current practices and opening up to more imaginative ways to create sustainable urban futures.

Introduction

Tales of transcendental urbanism

Eco-Cities: Urban Planning for the Future

> Massive developments proposed for the U.S., China and Abu Dhabi aim to reduce or even eliminate the environmental cost of city living. Treasure Island: a polluted military base is being transformed into a dense green neighbourhood . . . Dongtan: one of the earliest eco-cities announced may take longer than anticipated to complete . . . Masdar: Oil wealth fuels construction of an ecological city in the Arabian Desert.
> (*Scientific American*, special edition, October 2008)

The local energy revolution

> Big UK cities are looking at declaring independence from the National Grid and generating their own power. The carbon savings would be huge, but the barriers to switching off are formidable . . . Ken Livingstone's ambitious plan [is] to cut London's carbon emissions by 60% by 2025. 'To achieve the sort of carbon emission cuts that are necessary in London, we have to move from a society based on huge waste of energy to one that finds every available method to create energy without adding to carbon emissions,' says Livingstone.
> (*Guardian*, 6 December 2007)

C40 goals for UNFCCC Copenhagen 2009

> Cities and urban areas consume 75 per cent of the world's energy and produce up to 75 per cent of its greenhouse gas emissions. The battle to prevent catastrophic climate change will be won or lost in our cities. The C40 Large Cities Climate Leadership Group is a group of

> the world's leading cities committed to tackling climate change. We have already demonstrated leadership and concrete action on reducing greenhouse gas emissions, but on our own we can only go so far. In the run up to the United Nations Climate Change Conference in Copenhagen in December 2009, national governments must recognise the crucial role cities can play in delivering national targets and stopping climate change.
>
> (C40 www.c40cities.org/about/goals.jsp)

The relationships between cities and their energy, water, waste, food and communications infrastructure networks are changing in important and significant ways that challenge our conventional understanding of urban resource flows. Major claims are now being made – as the vignettes above illustrate – that new eco-cities can potentially 'eliminate' the environmental costs of cities, that big cities aspire to realize 'independence' from national energy grids and that effective global responses to the challenge of climate change will be 'won or lost' in cities. These claims are representative of a wider shift in thinking about how major cities can effectively anticipate and develop responses to a changing global ecological context in ways which allow them to maintain their existing economic and spatial growth ambitions. Whether 'cities' can really guarantee continued access to key resources, strategically protect 'themselves' from climate change and resource constraint and maintain high levels of interconnectivity is a critical issue for contemporary urbanism and, for us, the key issue that we address in this book.

We argue that these new aspirations and strategic concerns with resource flows is not a phenomenon that is happening naturally but is the product of specific social, economic, political and spatial processes, and that these changes have profound implications for the mutual organization of cities and infrastructure and consequently for the shape of future urbanism. Not taking these changes for granted requires that we analyse the politics and processes of the production, mediation, reception and consequences of visions of the future shape of the city and the infrastructure networks and resource flows that underpin its reproduction. Our view is that it is often particular coalitions of social interests, forms of knowledge and financial resources that are being used to develop and mobilize dominant visions of what the relationship between cities and their infrastructure networks should look like and then in attempting to translate these into material reality. We, therefore, subject the relationships between cities, socio-technical infrastructure networks and resource flows to critical scrutiny of their politics and processes. In doing this a primary concern is to ask: to what extent are visions of future urbanism socially progressive and concerned with building collective planetary ecological security? Or are they promoting the rebounding of divisible ecological security in relation to particular social groups and places predicated on new and often hidden interdependencies?

The reconfiguration of cities, socio-technical networks and resource flows are now clearly becoming increasingly strategic to cities and are being mutually intertwined and reorganized within the rubric of a new, extended focus on ecological security (see Pirages and Cousins 2005) that is layered over existing concerns with military and political security. New sets of systemic pressures – particularly in relation to combinations of issues related to climate change, resource constraint and their implications for the maintenance of economic growth – are impinging strategically on the mutual reorganization of cities and networks and underpin attempts to securitize resource flows. These pressures, though, are interpreted variably by different social groups in diverse cities. With this background in mind, what happens at the level of the city response to these pressures is clearly of significant consequence for the future shape of infrastructure networks, resource flows and the reproduction of cities.

World Cities and Climate Change develops a theoretical framework and the conceptual tools to critically interrogate these relationships and their consequences. The challenge here is not solely to understand the configurations of existing social interests that constitute the relationship between cities, socio-technical networks – utilities, regulators, urban political decision-makers and so on – and the organization of resource flows, but also the new social interests entering the field in response to these pressures. We argue that it is both within and between narrowly constituted coalitions of social interests and the world's most powerful cities that these pressures are most vigorously being responded to, through the development of emblematic and exemplary responses. Fundamentally, we seek to understand what it means for a particular class of world cities and associated coalitions of interests to become active in the (re-)organization of planetary ecological resources and what it means to attempt to secure 'their' ecological reproduction both in respect of particular world cities and networks of world cities.

This book develops a critical synthesis of a new style of infrastructural development that, we argue, is emerging as a particular response to the systemic pressures confronting world cities and their infrastructure networks. We are particularly concerned that these new fixes are being presented as emblematic and exemplary solutions that can be unproblematically mobilized, transferred and inserted into other urban contexts. Consequently, we will critically assess the relevance of these solutions developed in world cities to other urban contexts such as ordinary cities (Robinson 2006) and the mega cities of the global south, which are not fundamentally sites for the production of these fixes.

While we realize that addressing these issues comprehensively will require long-term studies and in-depth evaluation, our concern is to act now and ask critical and informed questions about this emerging style of urbanism. Powerful social interests are seeking to ensure that particular technical fixes, socio-technical configurations, and selected trajectories of urban development are normalized, routine, rolled-out and widely replicated across many urban contexts. There is very little fundamental questioning of this new logic and why it is emerging in the particular form that it is. For us, how history and critical analysis can help us to understand what is missing from these new configurations is crucial. By subjecting the claims of replication and transferability to scrutiny we can better understand what wider societal lessons we might learn from these initiatives. We can, furthermore, develop a better appreciation of what alternatives and options are being hidden or closed down by the focus on exemplary solutions and what opportunities for a different styles of urbanism based on collective planetary security are being missed. This book interrogates these issues in order to create a better informed and more sophisticated understanding of future urbanism in the context of a changing global environment.

Global urbanism in the anthropocene

Our motivation for writing this book was located in our efforts to understand important interrelationships between global urbanism and global ecological change and how they might challenge our comprehension of contemporary cities and the future of urbanism. In particular we wanted to construct a more relational and multifaceted appreciation of the complex relations between the significant social achievement of global urbanism – with more than 50 per cent of the world population living in urban areas – and wider recognition of a changing global ecology that is starting to reshape the conditions under which cities seek to guarantee continued growth and reproduction. We felt dissatisfaction with social visions that represented cities either as the root cause or as passive victims of global ecological change. Similarly, we were uneasy with the claims made by some cities that they could provide the exemplary or emblematic solutions to the challenges posed by climate change and resource constraint. We wanted to build a more nuanced and sophisticated understanding of contemporary urbanism and its future challenges through a set of issues that struck us as potentially critically significant in rethinking the future of urbanism.

Geologists have seriously argued that a new and distinctive geological epoch has begun which they call the anthropocene (see Zalasiewicz et al. 2008; see also Dalby 2007 and Figure I.1). It is proposed that human actions are the dominant process shaping planetary ecological change and that a measurable set of physical markers – these include disturbances to the carbon cycle and global temperature, ocean acidification, changes to sediment erosion and deposition, species' extinctions and so on – now characterize the contemporary geological period. Human activities are so pervasive and profound in their consequences that they are now altering the planet's ecological systems in ways that will reshape the global conditions in which humanity attempts to ensure its continued social and material reproduction. In line with this view, the United Nations Environment Programme (UNEP 2007) has concluded that human activities now 'dwarf' natural processes in reshaping global ecologies. The implications of such a shift are profound and are only beginning to be slowly understood and grappled with.

Figure I.1 Cover of *Geological Society of America*.
Source: GSA Today.

Recognition of the markers that characterize the anthropocene – which started in the early eighteenth century with the development of steam power – also corresponds with the profound social achievements of global urbanism and, in particular, the massive increase in urbanization and the proliferation of all cities globally. In 1900, 10 per cent of the world's population lived in cities, which 100 years later had risen to 50 per cent of the world's 6 billion inhabitants living in cities and by 2050 it is projected that nearly 70 per cent, of a global 10 billion population, will live in cities (United Nations Population Fund (UNFPA) 2007). Critically, the relationships between anthropocenic change and global urbanism are not coincidental and explicit connections are now being made between urbanization, changes in the biosphere and global ecological change.

Global urbanism is the dominant socio-economic process reshaping global ecology. It is doing so through the operation of the complex metalogistical systems that make the existence of cities possible (Luke 2003). All cities to differing degrees are embedded in water, waste, transport as well as energy, material and food networks that usually have national and international reach simply to provide the critical resources necessary to support the concept of urban life. Cities need to be understood as manufactured environments that are made possible by their interconnections with multiple infrastructure systems that tie them into logistical networks for the production and consumption of critical resources and commodities that make urban life possible. The concept of cities as metalogistical spaces captures this sense of contemporary urbanism as formulated through and constituted by dense concretions of infrastructural logistics. The prefix meta also then helps to view the city as an intermediary which sits as a site that anticipates, modifies and excretes the transformation of resources and materials and movement of goods and people. Consequently 'cities leave very destructive environmental footprints as their inhabitants reach out into markets around the world for material inputs to survive, but these transactions are also the root causes of global ecological decline' (Luke 2003: 158–159). Critically then any understanding of cities also has to take seriously the analysis of urbanized international and environmental relations of their resource flows. The social, technical and ecological organization of supporting such massive concentrations of people in huge agglomerations has become more distinct and challenging for how we live together.

In the contemporary period there is now recognition that industrialized systems, largely designed to support cities, and mostly controlled by organizations located in large world cities, have ecological effects that are beginning to change the global ecological context within which attempts are made to ensure the continued reproduction of cities. The paradox of the major social achievement of over half the world's population living in cities is that this has led to the claim that the ecological damage produced by the industrialized systems that support cities is leading to a form of planetary terracide as critical ecosystems are destroyed by human activity (see Reeves 2008). Cities will have to contend with a new atmosphere, changing oceans, different biodiversity, constrained resources and remade land masses with few warnings or obvious adaptive solutions. This means that a critical issue for twenty-first-century urbanism is how to understand and respond to the global ecological pressures that are produced by global urbanism and that fundamentally reshapes the material and resource context in which cities attempt to secure their continued reproduction.

While cities exist within a highly unified and integrated global space of capital flows, particular cities – and groups within cities – vary widely in their access to ecological resources. Highly energy intensive urban environments in the United States contrast with cities of the global south where millions do not have access to clean water, energy and basic telephones. The United States has only 4 per cent of the world's population but it generates almost 25 per cent of global carbon dioxide emissions (Soubbotina 2004). Americans' ability to tap into and control global ecosystems of fossil fuels mean that US cities are able to be far more spatially expansive and destructive than if they had to survive solely on the resources available in their national space. In an era where cities' ability to ensure their longer-term reproduction will be dependent on their capability to guarantee their ecological security and access to critical infrastructure and resources our concern is that there may be significant differences in the capabilities of cities to respond effectively to resource security and climate change.

Paradoxically as greater uncertainty and vulnerability characterizes global ecologies, globalization and the increasing role of markets and the private sector in the provision of infrastructure and resource flows have accelerated ecological damage and global ecological change. Processes of globalization and neoliberalism have produced a regulatory, policy and economic framework that has sought to rapidly accelerate the flows of resources, peoples and services into cities, the transformation of and value adding of flows and resources within those cities, and then the subsequent sale and the flows of the resultant commodities out of the cities. This has created more complex and intense flows of resource exchange and transformation, between locations scattered around the world, but has focused upon transformation in – and has been controlled by large world cities. Yet these processes are driven by a market logic with insufficient focus on the ecological and ethical implications of these enhanced, accelerated and liberalized resource flows. The dominant mode of development has been predicated on accelerating global flows between dispersed cities but with blindness to the materiality of these flows and their wider ecological and social consequences. What this means is that the city as a system works because of imported flows of energy and materials. Conventional socio-technical systems depend on drawing in vast resources of carbon-based energy sources such as coal, oil, natural gas and mobility that result in emissions, such as greenhouse gases, that threaten the viability of the biosphere.

However, a series of new pressures is now making these flows more visible to those staking a claim on world cities as they take strategic interest in the vulnerability and resilience of such flows. Coalitions that claim to represent cities have become increasingly interested in understanding how these material flows are constituted and how secure they are in the longer term. Central to this is the pushing of cities – through coalitions – into concerns about the vulnerability of critical infrastructure networks, not just in terms of the political and military security of networks located in cities but more in respect of the longer-term reliability and security of the critical resources that flow over infrastructure networks. Social interests in cities are increasingly active in developing an understanding of the long-term security of resources. As national states become mobilized around energy, water and food security, so large cities start to take a strategic interest in resource issues. These concerns translate into variable pressures in different cities and concern focuses on the urban implications of global ecological change, resource security and anticipating disruptions in production and consumption systems. As cities are major consumers of resources and producers of wastes and emissions then knowledge and understanding to ensure the strategic protection of the city needs to be developed. This is particularly critical in that it was these very resources that made the notion of rapid and dense urban agglomerations possible in the first place. Cheap oil and telecommunications made suburbanization possible, while controlling water supplies and waste disposal has been critical to urban growth and developing reliable food systems was fundamental to the creation of large mass urban populations. Increasingly the vulnerability of these resources in the longer term and the more immediate impacts of shocks – such as oil price rises – now mean a strategic concern with understanding questions around energy security, water security and food security is critical to large cities.

Producing *World Cities and Climate Change*

The issues and themes that we address in this book have been of longstanding intellectual interest and curiosity. Much of the empirical work that is detailed here was undertaken in the mid to late 2000s. It encompasses a range of world cities and eco-city developments but concentrates on responses being generated in London, New York, Tokyo, Shanghai, Melbourne and San Francisco and in the Thames Gateway, Dongtan and Treasure Island. In these contexts we have sought to deepen our understanding of infrastructural transitions and their relationships to world cities as responses to a set of systemic pressures – in relation to economic growth, climate change and resource constraint. More specifically we endeavoured to understand the visions of world cities and the systemic transitions envisaged in the organization and configuration of socio-technical infrastructures. This meant developing a richer understanding of not only these characterizations but also the processes through which they were produced, the field of multilevel governance relationships that produced them, how it was envisaged they were to be translated into a demonstrable materialization and what the consequences of this were.

This we did through a mix of methods that included engagement with these issues through secondary, grey literature. This was as many urban responses are unfolding within a temporal frame whereby there has been little academic literature produced on them. The grey literature included official literature, reports, strategy documents and promotional material of municipal authorities, mayors and city government departments as well as those of relevant multinational corporations and national governments. Our interest was in understanding in what ways these documents presented the relationships between cities and networks, who was doing so, why and what the consequences of this were.

We also subjected the propositions emerging from our reviews of secondary literature to critical scrutiny through a series of interviews with urban policymakers and decision-makers, utilities, regulators, national government officials and civil servants, architects and planners. We undertook over 60 interviews in research visits to London, New York, San Francisco, Tokyo, Melbourne and Shanghai. Through these interviews we sought to understand how cities are developing and endeavouring to enact strategic responses to a set of new pressures through the managed reconfiguration of socio-technical infrastructures both within metropolitan areas and externally through their relations at other scales.

In framing our argument we have emphasized the increasingly interrelated sets of pressures that are emerging. More specifically, this includes the promotion of economic growth in an era of globalization and neoliberalization, the geopolitical struggles around ecological resources, in a context where there are constraints around oil, natural gas and water, and where the carbon-based economic growth paradigm is under considerable strain given the consequences of climate change. We have done so from a position of commitment to the development of a theoretically informed understanding of not only these pressures but also an appreciation of the responses they elicit in the particular contexts of world cities, the relational production of these responses and the potential for these responses to be 'other'. This has necessitated the crafting of a theoretical approach which recognizes the different aspects and work to which it will be engaged with. While our argument acknowledges the 'pre-eminence' and centrality of world cities in the development of responses to the pressures and possibilities of urban ecological security we are also committed to exposing the political production of this 'obviousness'. This is important to us but not sufficient; we also wish to open up the possibility for alternatives in a debate where other options are given little oxygen.

Given these issues our initial focus is on world cities where, as we will demonstrate, it is social interests in relation to these cities that are among a pre-eminent class of cities developing dominant and obvious strategic responses to global ecological pressures. They are national and international exemplars where the emerging social visions of responses to climate change and resource constraint have recently, and are currently, being forged. Through our engagement with developments in these cities we are concerned to illustrate the particularity of the coalitions of social interests developing these social visions which are taking on an emblematic status and through which they are seeking to influence transitions in relations between cities, socio-technical networks and resource flows. We are particularly keen to begin the task of detailing the key political, social and material consequences of the unevenness of the highly political and narrowly relational production of these social visions. It is these consequences we are concerned with, both between and within cities. In line with our commitment to a critical but constructive engagement we are keen not only to illustrate the consequences of the dominant responses to these new pressures but also to provide for the possibility of alternatives to these responses.

The structure of the book

Our argument is structured in three parts with six chapters. Part I situates the book in wider literatures by focusing on the conceptual issues and challenges raised by different theoretical frameworks for understanding the changing relations between cities, infrastructure and resource flows. The part is divided into two chapters. Chapter 1 develops a critically informed, theoretical and conceptual approach to analyse and understand how relationships between cities, socio-technical infrastructure networks and resource flows are deeply political and uneven. The chapter examines the historically produced ideological shaping of these relationships and the prioritization of world cities within them. By mobilizing ideas of space the chapter explores how we can analyse the ways in which these relationships have been, are and can be governed. We develop a conceptual framework that focuses on the interactions of multiple levels or scales of governance and the material and symbolic struggle, negotiation and mutual development of cities *and* networks. Chapter 2 then characterizes the emergence of a new set of pressures and issues that are raised by the emerging sets of relationships between cities, climate change and resource constraint. We examine the relationships between different conceptions of security – geo-political, military, resources and so on – cities and their critical resource flows of mobility, water, energy and food. In particular we seek to understand how these relationships may frame a strategic reshaping of cities and key socio-technical systems. We then assess how these shifts interrelate with a wider set of questions about the economic reproduction of cities and the way in which ecological resources may become increasingly central to questions of competitiveness, innovation and employment.

Part II makes a shift away from theory and concepts to examine contemporary urban responses to global ecological change. This part has two chapters. Chapter 3 presents a framework for understanding how cities are selectively seeking to rebundle together fragmented and competitive infrastructures into new networked ecologies. We then explore different types of response in more detail focusing on critical examples demonstrating how responses are developing across many different urban contexts. Finally the chapter presents a typology of responses to new pressures. Chapter 4 focuses on the responses developed in world cities characterized as secure urbanism and resilient infrastructure (SURI). We address the issue of how it is claimed the economic and ecological well-being of world cities can be secured in a context of interrelationships of rapidly growing population, demand for resources and resource constraints, climate change and intense competition for economic activity and jobs. We argue that coalitions of interests in relation to world cities have an emerging and increasingly influential role in developing responses to the challenges of resource constraint and climate change where world cities are being positioned as the contexts for action. Our concern in this chapter is that the strategic reorientation to resource constraint and climate change is leading to the development of new styles of infrastructure development that privilege particular spatial and socio-technical configurations of infrastructure around strategic protection, building autonomy and developing new global urban agglomerations.

Part III addresses the wider implications and critical limits to bounded urban responses. We set out our own critical concerns that SURI represents a narrowly conceived, technocratic, under socialized, overly securitized and competitive response that seeks to build bounded archipelagos of networked ecologies focused on world cities. This part is divided into two chapters. Chapter 5 considers the wider implications of urban responses through SURI and replicant eco-cities. Specifically we argue that this logic reinforces new strategies of economic accumulation within the existing urban hierarchy rather than signalling a more progressive kind of politics that may challenge existing urban hierarchies. We examine six critical sets of issues that, for us, raise worrying concerns about the wider implications of SURI. Chapter 6 sets out our concluding position on the new styles of response, encapsulated by SURI, as being fundamental to the production of privileged spaces of securitized ecological urbanism. We conclude that there is compelling evidence that 'urban ecological security' (UES) has become a strategic issue for world cities as they seek to build divisible ecological security at the metropolitan scale. However we then open up the possibilities for intermediary spaces of response to the pressures of UES by developing a framework for constructing urban responses that are more equitable in their constitution and their consequences than SURI.

PART I

Theory, Concepts and Issues

1

The politics of governing cities, infrastructures and resource flows

Introduction

The latest phase of neoliberal globalization has seen the re-emergence of cities and regions as units of economic organization where cities and regions are propounded as motors of economic growth. The global urban growth agenda that has been dominant since the 1980s has taken place at the same time as – and indeed encouraged – the growth of urban populations to the point where the majority of the world's population now live in cities (UNFPA 2007). Cities thus not only are promoted as central to economic growth strategies but also are often centres of expanding human populations and associated places of production and consumption of resources and associated emissions. In western industrialized cities the very natural resources that underpin neoliberal forms of growth and the infrastructures that support the flows of these resources have been subject to spatially variable reorganization; reorganization that is a contributor to and consequence of the liberalization, deregulation, privatization and marketization of energy, water, waste and transportation systems.

From the late 1980s and through the 1990s the dominant view of urban and regional development, propounded through ecological modernization and sustainable development discourses, was that the relationships between continued economic growth, ecological protection and equitable social development could be managed. Yet sustainable development and ecological modernization, in mediating urban and regional strategies, have been seen to be appropriated and incorporated into neoliberal accumulation strategies (see Brand 2007).

Increasingly the extent to which urban and regional strategies are able to demonstrate that they hold together economic, ecological and social equity goals are additionally being questioned by a series of wider geopolitical pressures. Patterns and flows of resources, the organization of their production and consumption, are being challenged through what is increasingly being characterized as resource constraint – the 'peaking' of oil and gas, water scarcity and the environmental problems created by waste and landfill – and also through the spatial enablements and constraints afforded by the cascading of internationally and nationally negotiated climate change priorities, targets and associated carbon regulations. These developments are creating pressures on the ways in which socio-technical infrastructure systems, many of which were initially designed in the late nineteenth and early twentieth centuries, and resource flows are organized in western contexts.

This is important because cities are increasingly being promoted, by a wide variety of urban, national, business and environmental decision-makers, not only as economic motors but also as significant sites in the production of carbon emissions and the consumption of natural resources. Yet it is these new pressures that the city – through the coalitions of social interests that claim to speak on its behalf – must increasingly be aware of and be able to effectively respond to. This is of strategic importance because the resources that underpin contemporary urban economies are increasingly constrained. The emissions that are produced through urban economic activities are also increasingly being bounced back onto cities through the spatialization of carbon regulation and the consequences of climate change-related extreme weather events. These consequences are likely to be keenly felt in cities, with their concentrations of population, assets and often coastal locations. This means that the coalitions of social and institutional interests that claim to speak on behalf of the city must develop responses that are mediated through existing structures, economic, ecological flows and governance frameworks, which then informs attempts to reconfigure relationships between cities, socio-technical infrastructures and the organization of resource flows.

Yet, global urbanism, as a relational territorial process of reshaping cities, infrastructures and resource flows, means that the privileging, prioritization and success of some cities or populations within cities and the deprivation, neglect and stagnation of other cities should not be seen as disconnected but are products of a shared process (Luke 2006). That is to say that what happens within and between those cities and the populations which are seen to benefit from the reshaping of city–infrastructure–resource relationships cannot be divorced from interdependent consequences for other cities and populations.

Our fundamental preoccupation in this chapter is, therefore, to develop a theoretical and conceptual understanding of the dynamic relationships between cities, the socio-technical organization of infrastructure and myriad resource flows. We do this in a wider context of political, socio-economic and institutional shifts that have taken place since the late 1960s or so, that have been characterized as neoliberalism. Importantly, we seek to build on these debates in the face of new emerging, and on the face of it apparently contradictory, pressures facing cities and the organization and provision of resource flows underpinning particular spatial manifestations of economic, social and ecological reproduction.

To do this we critically theorize and conceptualize relationships between cities, infrastructure and resource flows as spatially, politically, institutionally and relationally produced. In doing this we articulate a view of cities as dynamic social spaces underpinned by infrastructural unevenness and a variable provision of, access to and recipients of resource flows. This is not naturally produced but is the consequence of governance coalitions that have the potential to embody a variable participation of social interests – from urban government, through utilities, regulators, business, consumers, citizens and so on.

In doing this the chapter has eight sections. First, we broadly characterize the organization of relationships between cities, infrastructures and resources in western, industrialized countries from the Second World War to the current time. Second, we examine the role of space and politics in governing relationships between cities, infrastructures and resource flows. Third, we assess the changing role of the nation-state and the ways in which it informs (uneven) urbanization. Fourth, we examine the prioritization of urban economic governance, the promotion of urban competitiveness and growth and the central role for governance coalitions in relation to world cities, infrastructure and resource flows. Fifth, we open up understandings of urban economic governance to rethink relationships between city, infrastructure and resource flows. Sixth, we raise the prospect of an emerging set of new pressures to the ways in which cities, infrastructure and resource flows are organized under conditions of neoliberal urbanism. Seventh, we devise a theoretical and conceptual approach to reflect on the reorganization of city, infrastructure and resource flows as predicated on tensions between transformative reconfiguration and obduracy from historically produced coalitions. Finally we present conclusions.

The organization of post-Keynesian urban infrastructures

Since the late 1970s we have seen a challenge not only to the hegemony of discourses of infrastructures as administrative service delivery mechanisms but also to the hegemony of the Keynesian-Fordist nation-state and its relationships with its subnational territories. In this respect it is possible to broadly characterize structural changes in the socio-technical organization of infrastructure provision in relation primarily to western, industrialized countries. We contrast two periods of infrastructure provision – nationalized infrastructure provision and spatially differentiated infrastructure provision – to highlight the importance of understanding structural change in shaping the relationships between cities, socio-technical infrastructures and the production and consumption of resource flows. In doing this we recognize that often these characterizations serve as ideal types, that there are no sharp disjunctures between these phases of provision and that, in many ways, there continue to be struggles between these competing ideologies.

Nationalized infrastructure provision, from around 1945 until the 1980s, was a product of the post-war period which created public monopolies that focused on the integration of national economic space through standardized infrastructures. The dominant philosophy saw a key role for central government intervention in economic planning generally and, more specifically, the use of public corporations in the provision and management of infrastructure. In nationalizing networked provision, existing local networks were extended and linked into regional and national grids that both covered cities and interconnected urban systems. This was underlain by vast capital investment in rolling out a standardized, ubiquitous and taken for granted national infrastructure.

Urban policymakers and planners were able to assume that infrastructure services would be largely provided where and when they were needed, leaving the specific details of implementation to engineers and network operators. Infrastructures were managed as public service monopolies in the broader public interest, which entailed a commitment to widespread provision of broadly comparable – standardized – networked services largely regardless of geography. Consequently there were not significant variations in the cost and quality of infrastructure between cities and regions. A key consequence was that individual cities and regions had relatively little impact on the process of network provision and management.

Since the 1980s the nationalized approach to infrastructure provision has encountered a series of ideological, political, regulatory, technological, social and managerial challenges. These pressures have questioned the previously taken for granted view of infrastructure being largely about the administrative provision of technical networked services. The centralized, nationalized supply logic has been challenged by a neoliberal logic of privatization and liberalization in response to fundamental structural changes in the global economy.

Consequently, from the early 1980s onwards, nationalization was increasingly questioned as a general principle and, more specifically, in terms of government attempts to improve the economic efficiency of the infrastructure sector. Large, centralized, nationalized providers were seen in some quarters as being inefficient, unaccountable, stagnant and uncompetitive and the supply logic of nationalized provision, built on a perception of plentiful supplies of resources, was also questioned following the oil crises of the 1970s. This, when added to rising concerns about the environmental consequences of supply-led growth and social resistance to new infrastructure, served to question the role of public, standardized, universalized provision of networked services.

Systematic processes of privatization and liberalization reshaped the institutional structures of infrastructures and technological organization of services. Integrated infrastructure networks were unbundled into competitive and monopolistic segments that provided infrastructure through a range of different services, customized around the requirements of particular customers and places. The removal of cross subsidies from large to smaller users weakened providers' commitment to standardized services as they offered different packages of services to an increasingly segmented market.

Infrastructure provision, in contrast to large centralized providers, shifted towards a mix of competing private companies encompassing old utility providers and new entrants predicated on 'competition' between infrastructures across a range of services. The notion of an underlying national interest was challenged by a commercial logic through which the possibility for spatial differentiation in terms of networked provision was created. This was underpinned not only by the politics of privatization and liberalization but also by their close alignment with the availability and development of new networked technologies. This bringing together of privatization, liberalization, a range of competing interests and new technologies provided a context within which infrastructure networks experienced pressures to reconfigure. With the shift from nationalized provision, this opened up questions of how we understand networked provision as a complex and uneven patchwork where utilities are being developed at the local, regional, national and international scales.

The key point is that cities now operate in an increasingly complex infrastructural environment. This is characterized by a patchwork of providers, often privatized, operating in liberalized regulated markets, where regulators' roles in setting prices are frequently in tension or conflict with spatially specific plans to spend and invest in infrastructure to enhance environmental performance. Infrastructure is increasingly characterized by differentiated styles of service provision that vary across users and place. Usually cities do not have an explicit role in shaping the priorities and objectives of network providers according to place-specific territorial objectives. Consequently, governing the relationships between city, infrastructure systems and resource flows where the boundaries of each is often less than clear, is complex.

Space, politics and governing cities, infrastructure and resource flows

Commonly when we think about the relationships between cities, infrastructures and resource flows we do so in very particular ways which address the impact of infrastructures (or a lack of them) on places, the provision of networked services for places, network failures in relation to places, shortages or uncertainties in resource provision and so on. Frequently the plans, images, maps and schemes that represent attempts to rethink and reimagine relationships between cities and different networks do so, partly out of necessity, in ways that simplify and reduce it to a fixed absolute space (Harvey 2006).

Yet, these ways of thinking about these relationships say little about why they are organized the way they are and how the representation of absolute space is relationally and relatively produced (Massey 2005; Harvey 2006). Much of the relational and relative politics, struggle, social interests and processes of negotiation underpinning not only such representations but also the realities they seek to represent are lost or obscured. The point is that the relationship between cities, infrastructure and the organization of resources that are produced, consumed and flow through cities is not unproblematic and is deeply political. Governing this politics requires new techniques of government and governance through the interactions and negotiations of a range of institutional and social interests and understanding the constitution of such governance coalitions and their capabilities to act. At one level the struggle is concerned with relationships between cities, infrastructures and resource flows and the balance and struggle between maintaining the way things are, why they are that way, but at another level how they might be reconceived.

Representational space (Massey 2005) allows views of complex and deeply political relationships between infrastructures, resource flows and places to be articulated and framed in highly exclusionary ways. Often this results in an equation of space and representation where both the spatial and meaning are fixed, closed, static and discrete and in doing so space obliterates processual time to the privileging of instants. It also excludes a multiplicity of coexisting narratives and potential narratives. Space as representation in this way is representative of a view of place that already 'exists', and through representation is re-enforced, rather than constitutive of what space and place can become. In this way, fixity in representations of space frames an ideological closure through imagining, defining and attributing particular meaning to places (Massey 2005).

Seeing space as actively, relationally and relatively produced (Massey 2005; Harvey 2006) means that we can understand attempts to define and categorize cities not only in relation to existing and absolute representations of the city but also through the ways in which the field of interrelationships of institutions and social interests seek to reconstitute an identity for the city. This is not to view how we understand place as reducible to a narrow localism but to see it as produced through coalitions of multilevel governance interests, institutions, pressures, forms of knowledge, technologies and resources in the widest sense of that term. The issue is how this is done and we will return to this later in this chapter.

Not only are spaces relationally produced but also they are understood in the way they are organized relative to each other. Spaces are produced relationally and, in the competitive context of neoliberal globalization, coalitions of interests have engaged in a 'race' vis-à-vis each other to represent categorizations and identities of success. This has led to attempts to replicate and ape what are represented as successful cities and places and has seen the fixes developed in particular spaces and places represented as replicable between contexts. Relative space allows us to puncture the veneer of the unitary spatial representation of the city and to uncover the multiplicity of varying spatial configurations within cities. The result is the exposure of the oxymoron of seeking to replicate spatial distinctiveness.

Relationships between different notions of space are significant. The ideological closure that spatial fixity produces can be opened up through thinking relationally and relatively about space and place and the possibilities for multiple spatial imaginations to flourish in ways which deny fixity and acknowledge the dynamism involved in making and remaking space. This is the tension between a narrow politics, usually of protecting the status quo and of a transformative understanding of space. It is related to the process of becoming rather than of staking a narrow claim for meaning and categorization as to what already exists. Space is not a container for the representation and persuasive mobilization of the desirability of already constituted identities nor is it completely closed but it is a locus for constituting a view of what the city could and should become. In this way our understanding of space is critical to understanding urban processes.

Viewing each of the component aspects of this triumvirate – cities, infrastructure and resource flows – not as absolute spaces or containers but through their relational and their relative production vis-à-vis other spaces provides the basis of a helpful way of thinking about the mutual and political production of cities, infrastructure networks and resource flows. Thinking relationally and relatively about the mutual spatial production of cities, infrastructure and resource flows allows us to think about the way we conceive of and understand cities and the way we understand the infrastructure systems that mediate relationships between resource flows and cities. Relationally this means thinking in terms of the networks or fields of institutions and social interests that coalesce to stake a claim to speak on behalf of the city.

This view sees the city as relationally and dynamically constituted from networks of asymmetric and unfolding relationships that include and exclude. These are asymmetric not only in the sense of the local social interests who bring different forms of financial resources, expertise and influence to networks but also in respect of wider multilevel governance relationships where the relationship of national state and/or supranational political institutions to territorial social networks are structured in highly variable ways. To put this another way, different levels of authority work on the same territory that is subject to constant reinterpretation. Not only is territory subject to constant reinterpretation from multiple levels of authority and positions but also it is central to the struggle and negotiation for classification. Coalitions of multiple social interests – the 'who' of the city – often with varying expectations seek to shape its future orientation and what it should become.

The competition state, distributed governance and neoliberal urbanism

There are significant geometries of power associated with coalitions of interests. This is underpinned by the promotion of liberalization at a variety of levels of governance, from the World Bank and the International Monetary Fund (IMF), down through, for example, the supranational European Union (EU), to regional development agencies and also to bring a variety of the ever increasing International Non-Governmental Organizations (INGOs) and also Quasi-Autonomous Non-Governmental Organizations

(Quangos) at both the national and the subnational level into the realms of governance. This has not only meant the widespread diffusion of broadly shared values but also resulted in centres of power and authority that have been distributed and often privatized. That said, it is possible and necessary here to characterize the shifting role of the nation-state in terms of an emerging spatial differentiation in relationships with the subnational.

This is likely to vary between coalitions and places but the changing role of national states from welfare towards competition states and the spatial strategic selectivity upon which such a shift is predicated is critical to understanding struggles to reproduce places (Jessop 2007). Neoliberal globalization, although frequently promoted as natural and witnessing the end of, or at least a weakening of national states, is politically and ideologically driven. The processes of restructuring that reproduce neoliberalization have paradoxically often been promoted by national states rolling back Keynesian Welfare states through privatization, deregulation and a retreat from welfare provision and at the same time rolling forward national states through a huge number of deregulatory and privatization measures, which have added to the sense that there is no alternative to more market and less state.

The issue then is how to theorize the role of the nation-state, its key features and implications for cities. One compelling view is that we have seen a shift away from forms (in the plural) of the Keynesian Welfare state towards (forms of) the Schumpeterian Competition State (SCS) with a 'concern with technological change, innovation and enterprise and its attempt to develop new techniques of government and governance to these ends' and where 'the crucial point is that state action is required to encourage the development of new core technologies and their application to as wide a range of activities as possible to promote competitiveness' (Jessop 2002: 96 and 261).

The crucial point for us is that aligned to the role of the SCS as promoting labour, currency, trade and investment liberalization is its advocacy of 'national or regional capitalisms and appropriate conditions for their global spread'. In doing so it promotes uneven development through interurban and interregional competition. This is part of strategies to appropriate and fix hyper-mobile capital in their national economic space, in a context where knowledge has become fundamental to the conduct of economic activity (Jessop 2002: 138; 2007: 192–193). The role of the state is one of seeking to manage the tensions and contradictions between mobile, often global capital and the territorial and political issues and implications of this. What this means is not the end of the nation-state but a degree of hollowing out through a reorientation in the functions of the state and the uneven shifting of some tasks to both the supranational level and to the local and in doing so the promotion of local, urban and regional forms of governance.

Furthermore, there is a contradiction between increased competition based on short-term economic calculations and the long-termism required in terms of the development of socio-technical infrastructure systems. This is a contradiction between the economy seen as a pure space of flows and as a territorially, socially embedded system. Seeking to encourage competition in an era of multilevel governance means that representatives of different places, businesses, and institutions at different levels of governance increasingly need to find ways of coalescing to act but do so from different positions and with varying motivations. Governance coalitions create and respond to the promotion of increasing international economic competitiveness through encouraging innovation. Central to this at different levels and across various social interests is knowledge which needs to be effectively mediated to inform new forms of collective action.

The critical point for us here is that the systematic unpicking of Keynesian-Fordist states, with their view of state space as relatively homogenized and universalized, is subject to reorganization through sociopolitical struggles that produce and reproduce lattices of spatial configurations. Within this wider context, urban regions have become significant sites of state, institutional and spatial restructuring (Brenner 2004). The crises of Fordism and the shift to new growth strategies based on urban entrepreneurialism has entailed a shift from national spatial planning to processes of uneven development where 'social, political and economic processes under capitalism are not distributed uniformly or homogenously across the earth's surface, but are always organized within distinct sociospatial configurations' (Brenner 2004: 12–13). It is particularly in city-regions that these processes have been unfolding.

In relation to such sites the dominant organizing principles have been economic competitiveness, entrepreneurialism, innovation and so on. The discourse of urban entrepreneurialism (Harvey 1989) has become pervasive. Entrepreneurial cities engage in self-marketing, constantly constructing and reconstructing images to sell themselves as a means of attracting capital, people and ideas in the space of flows (Castells 2000). Success for cities in the competitive race is fundamentally linked to their economic viability and growth. This informs a constant search for innovation to be and remain competitive predicated in the first instance on the ideology of mobile capital rather than any more specific sets of local priorities. Given the competitive underpinnings of the race, images of such urban success are constantly produced and communicated not only as a means of positioning cities vis-à-vis each other as relative spaces but also as a mode of coercion to force cities to constantly act rather than to drop out of the race.

This sees a shift in the governing of cities from a redistributive, managerial function to an entrepreneurial role with a central role for local growth coalitions (Logan and Molotch 1987) in constructing a distinctive image or representation of the city as an attractive place. This also involves a governance focus on public–private partnerships in the cultivation of, for example, waterfront developments, mega-events and so on, where there may also be national government assistance to encourage foreign direct investment. This coincides with but is not dependent on taking a view of the city as one where relationships are constantly reworked and the city redefined.

The dominance of economistic urban governance and world cities

The dominance of economistic thinking in urban governance is a particularly important focus. The social interests which coalesce around urban entrepreneurialism in some sense represent specific cities in particular ways. Much of the recent literature concerned with the governance of urban transformation and the active role of cities has focused on economic governance and the construction of growth coalitions or regimes of local political elites and business (Logan and Molotch 1987). In this respect, the growth machine thesis argues for the centrality of cities as engines of economic development for business interests and, in doing so, provides a rationale for how capital flows and investments are appropriated to drive urban growth. This involves a central role for urban policy and politics and the mobilization of growth through urban growth coalitions of government and local growth elites, including local politicians, the local media, quasi-public agencies and auxiliary players (Logan and Molotch 1987; Harding 1995). In starting from the view of cities as engines of economic development, but with a fundamental role for business and political interests, ideologically this closes down views that question economic growth.

The associated urban regime theory (URT), with a background in the United States, focuses on the emergence of coalitions that arise in cities – coalitions that are largely informal but stable and coherent and coalesce around a broad commitment to, for example, economic growth in a particular sector, albeit often for a variety of different motivations. Although the growth machine highlights the potential for urban growth coalitions the hegemonic dominance of urban entrepreneurialism is not unproblematically assumed but is seen as one possible coalition among many but one which has assumed hegemonic dominance around corporate interests.

Within these parameters, URT then allows understanding of the ways in which different public, private and civil society coalitions in urban contexts cooperate, inform and produce responses to contemporary sets of issues and challenges. Resonating with neo-pluralist arguments, regime theory highlights that in the context of global interdependencies control in urban contexts is distributed. It is, thus, the construction of coalitions and the forms of power and the ways in which they are mobilized which are of interest but these regimes may look quite diverse in relation to different cities and coalitions constituted by variable sets – and weightings – of political, business and civil society interests. The form of regimes needs to be understood not only in terms of these endogenous coalitions but also in respect of different sets of challenges and issues particular cities face, the historical institutional contexts and arrangements within cities and also scalar political arrangements which can enable and constrain the possibilities for different regimes.

Yet, with this understanding of the building of growth coalitions and regimes in mind it is still a particular class of city that is prioritized in the era of neoliberal globalization. The re-emergence of the urban and regional in contemporary globalization is not equitable in its infrastructural manifestations and consequences. Urban growth regimes are predicated on new techniques of government and governance in attempts to enact competition-based change where these developing sets of relationships are situated in various institutional contexts. In relation to urban infrastructure this means that the competitive governance of economic activity becomes bound up with the governance of socio-technical infrastructure systems in very complex ways where boundaries between urban growth regimes and socio-technical regimes or systems are semi-permeable.

Attempts to enact economic innovation through infrastructural innovation have produced a situation where seeking to be a first mover city means acting competitively. Yet, one of the motivations, in a competitive race, for doing so is to set the parameters for those following and, in doing so, seeking to coerce their future actions through exemplifying what that future should look like while at the same time the illustration of that future is being seen in relation to a particular city.

It should come as no surprise that an emblematic feature of globalization and urban governance are networks between large cities and their particular positions in a division of labour in an interconnected world system. These have been characterized as world cities (Friedmann 1986) or global cities (King 1990; Sassen 2001), where infrastructures of transport, financial markets and circuits of financial services more broadly and information and communications technologies constitute connections between and flows through world cities predicated on not only technological architectures, flows of data, people and goods but also a series of multilevel political and economic relationships between political and economic elites. That means that they are a critical focus for understanding urban and infrastructural transformation. It has long been recognized that world cities play a critical role in a wider territorialized division of labour where they are key economic, financial and governmental command and control points. In other words:

> Cities are central to neoliberal globalisation. . . . The competition between them is both product and support of the neoliberal agenda. And in certain cities (those we call world cities, or global cities) is concentrated the institutional and cultural infrastructure that is key to all of this.
>
> (Massey 2007: 9)

Researchers have sought to understand the differential positioning of world cities against each other and with respect to cities further down the urban hierarchy according to their economic functions and connectivity (Beaverstock et al. 1999; Sassen 2001) and where cities figure within neoliberal globalization in very diverse ways.

A critical concern for national states has often been the development of policies and priorities that protect and enhance the role of *their* world cities, where such priorities are justified within the national interest because of competitive pressures from other world cities. Consequently we argue that there is significant conceptual and empirical evidence that world cities occupy a different position with respect to the organization of socio-technical infrastructure systems.

Connectivity and relationality constitute the underpinnings of power relationships between world cities, where this connectivity informs the competitive race between world cities to attract investment, labour and technological innovations. In this respect, infrastructural innovation in the context of world cities is often presented as exemplary, not only in underpinning the aspirations of world cities to attract investment, labour and technological innovations but also in framing the sorts of possible strategies assumed to be available to other cities. That is to say, frequently world cities, through coalitions of urban and national governance elites and business, are represented as *the* places where responses with a wide and transferable applicability to other cities are developed. This dominance of world cities is fundamental to arguments made for the centrality of cities as engines of economic development for business interests and, in doing so, providing a rationale for how capital flows and investments are appropriated 'entrepreneurially' to drive urban growth.

World city theories are not without critics and critical engagement (see Abu-Lughod 1991; Brenner and Keil 2006). The danger in the contemporary phase of global urbanism is in overplaying the role of physical and technological infrastructures, or a technological determinism (Smith and Marx 1994), which focuses on world cities as nodes and ignores the neoliberal political and ideological institutional shifts around privatization and deregulation which have encouraged the growth of financial, information and knowledge services and infrastructures to support this. Urban infrastructures do not solely impact on cities and their populations nor do they develop naturally. Relationships between cities and infrastructures can be seen as socio-technical (Hughes 1987) and not, thus, be understood through infrastructural technologies impacting on the city but through the co-construction of city and infrastructure.

They are socio-technical in that technologies and technological functions co-evolve with social functions and social interests where technological development is seen to be shaped and potentially shaped by a broad constituency of not only technologists and engineers but also policymakers, business interests, NGOs, consumers and so on. It is the interrelationships of these social interests, through regulations, policy priorities, consumption patterns, investment decisions, among other things, that hold together to stabilize socio-technical regimes and their existing trajectories (Geels and Schot 2007). Thus, the infrastructure systems – that provide water and power to cities – are not necessarily bounded at the level of the city and encompass social interests and institutions across multiple levels of governance. When set alongside the multilevel governance of urban growth coalitions this highlights the many actors, institutions and centres of power that produce the city from within and beyond its administrative boundaries. A key consequence of this is not to view world cities as a single unified category but to recognize that it is more accurate for characterizations to encompass the variability and differential pathways of global city formation (Olds and Yeung 2004) and the socio-technical systems that support them.

Opening up urban economic governance: rethinking city, infrastructure and resource flows

Although scalar approaches have been utilized in addressing urban economic governance, they have been relatively neglected by the literature around urban environmental governance, where space and scale are usually unquestioned (Bulkeley 2005). Environmental governance often reflects a cascade approach (Bulkeley and Betsill 2003) to political scale where there is a hierarchical and vertical set of relationships predicated on relationships between scales which often lack dynamism. An emphasis on inter-nation-state negotiations and their production of international environmental agreements is central to cascade approaches. The nation-state is characterized by the singularity of its position, thereby often neglecting a diverse variety of opinion and, in many regards, is seen as a black box. The negotiation of these agreements, subsequently sees a conferring of the role of ensuring (or not) compliance on the nation-state with the key role of the subnational as the scale at which environmental obligations will be met. Cascade approaches again view the local level as largely a black box.

The view of the role of the nation-state as encompassing a relatively unitary position often neglects the wide variety of different positions within national contexts. At the same time this – and the view of a black-boxed local – reinforces relatively fixed notions of boundaries and the interplay of different scales of political activity. Through viewing the local as container for the passing down and achievement of international agreements this overplays the stability of the governing of nation-states and underplays the possibilities for local forms of agency and spatial differentiation (Bulkeley and Betsill 2003: 15–18). These arguments about the hollowing out of the state, the re-emergence of the subnational and the ways in which scalar configurations of new forms of governance are produced have largely been missing from environmental governance debates (for notable examples to the contrary, see Bulkeley and Betsill 2003; Keil 2003; Desfor and Keil 2004; Luke 2008).

The prevalence of an academic concern with a policy discourse of urban economic competitiveness, lionizing narrow notions of entrepreneurialism, innovation and competition and its scalar construction has often resulted in a characterization which separates notions of economy and environment, frequently through characterizing environmental questions in terms of their constraints on economic growth. This is informed by an ontological separation of nature and society rather than viewing nature and society as mutually constitutive (Desfor and Keil 2004).

The predominant strands of world cities' literature with their narrow economic framing and concerns, exemplified through a small number of pre-eminent world cities, have often eschewed a critical analysis of and engagement with debates that seek to rethink the relationships not only between the global and the urban but also between nature and the city. Often world cities' approaches have underplayed or even ignored ecological and local political processes informing the reproduction of world cities (see Keil 1995). In doing so, the specificity and distinctiveness of urban ecologies and the particular ways in which local strategies and responses to globalized urbanization – predicated on local histories and experiences and locally embedded power relations and negotiations – are constituted are undervalued. World cities are constituted through processes where the relatively small number of pre-eminent world cities should be understood not only in terms of the (often private) networks of and relationships between these cities but also within the wider context of the pervasive processes of globalized urbanization. The reproduction of this particular class of world city is predicated on the worldwide webs of logistics flows where these cities are interdependent on human and non-human relationships enmeshed across the system of global cities conceived more broadly. Viewing the small number of pre-eminent world cities in relation to these broader processes provides the possibility to open up to scrutiny the privatized ecology of these cities and the potential for a 'highly public ecology, which must be repoliticized, resocialized, and relocalized by environmental activists in many everyday struggles' (Luke 2006: 277–278).

What this illustrates is that frequently questions of economic and environmental governance are not considered in an interrelated and integrated way in relation to cities and infrastructures. We know that dualisms of economy and environment, technology and society, global and local are questioned and contested, in that: 'Urban-nature relations are now increasingly constituted at various scales of the globalization process as natural relations and urban social relations are produced through complex processes of "glocalization" and entangled in myriad flows of capital, things and people' (Keil 2003: 729). This resonates with the arguments of urban political ecology (UPE) (Keil 2003; Swyngedouw and Heynen 2003; Heynen et al. 2006) which, with its view of the urban as simultaneously local and global and human and non-human, draws together aspects of political economy and ecological approaches, affording a focus on issues of the scalar, material and discursive construction, negotiation and contestation of cities.

Nature is not seen as external to the social world but as produced in entangled material, social, political and economic relationships that, in capitalist cities, takes primarily the social form of commodities. The circulation of flows of these relationships, resources and artefacts and their metabolic processing are central concepts and in the transformation of both physical and social environments inform the production of urban contexts (see Heynen et al. 2006). In this view, urbanization is predicated not on a separation of human life from nature but instead through a process by which new and more complex relationships of society and nature are created (Keil 2003). Consequently: 'Metabolic circulation, then, is the socially mediated process of environmental, including technological, transformation and transconfiguration, through which all manner of "agents" are mobilized, attached, collectivized, and networked' (Swyngedouw 2006: 33). In doing so social relations shape natural environments and vice versa and there is a mutual transformation of society and nature that produces and reproduces social and physical contexts, in our case, the city.

In the organization of energy, water, waste and waste water and transport systems, significant flows and circulation of natural resources are metabolized in ways which have variable distributional consequences and risks. The particularity of these consequences and risks relate to specific social, political, cultural and economic conditions and relationships where flows are mediated, regulated and controlled and the governance of these processes are contested and subject to politically motivated redefinitions. Processes of reproduction are not episodic but are ongoing and continuous consequences of metabolic circulations encompassing entangled relations of a multiplicity of 'agents', material and symbolic, physical and human operating in processes and networks organized across different global, national, regional and local scales of activity. What we can now see is that governance coalitions that are often organized in terms of the economic competitiveness race also appropriate nature, environment and technical artefacts that produce particular and variable effects for different places and social groups.

'New' pressures on the organization of cities, infrastructures and resource flows

The organization of cities, infrastructures and resource flows and the relationships between them, as we have seen, are complex, deeply political and rooted in ideological struggle, governed through inter-institutional negotiations often mediated through notions of competitiveness, entrepreneurialism and innovation in particular places. In this respect infrastructure innovation has been utilized to support place-based competition and also, through the development of services, to contribute to the creation of place-based competition.

The reproduction of these relationships, mediated through a neoliberal discourse of competition, innovation, liberalization and so on, is increasingly encountering a further set of strongly emerging pressures. Market-based economic activity and growth remains a fundamental mediator of relationships between places and infrastructure but increasingly serious consideration in organizing these relationships is being paid to the challenges posed by climate change, both in terms of anticipating its 'impacts' and also the strategic direction of national state actors in regulating and disciplining spaces. Furthermore, the geopolitics of oil, natural gas and water and also waste management questions – again with both geopolitical and more regionalized consequences – are contributing to a wider set of pressures and debates around resource management and, in particular the ways in which reconfiguring systems and the relationships between systems and places can effectively address resource constraints.

Increasingly the language of threats, risks, dependencies, vulnerabilities, impacts and security are being mobilized to confront the current organizations of cities and infrastructures in ways which encompass issues related to resource constraint and climate change. More tangibly this includes in respect of energy a dependency of large parts of the western world on the geopolitics of the supply of oil and natural gas from volatile regions and in particular the – contested – peaking of global oil production. There are issues here in terms of the long-term volumes of oil and natural gas that are available and the short and medium-term political organization of this through organizations – such as OPEC for example – that organize the regulation and production of oil. One consequence of this is pressure for many, particularly, western economies, to diversify the type and sources of energy production. A critical issue is the timeframe – likely to stretch over a couple of decades – of reconfiguring domestic energy supplies.

Not only is time an issue in respect of how long it would take to reconfigure energy systems based on fossil fuels (oil, natural gas and coal) but also it is in respect of constraints of resources – when oil and natural gas reserves, for example, will be exhausted is a critical issue as is when the impacts of climate change are to be felt. This is important not only in respect of assumptions and scenarios but also in terms of varying perceptions as to the availability of resources and the consequences of climate change. Arguments around the scarcity of resources are not altogether new. Debates around the earth's limits to growth have been lively since the 1970s (Meadows et al. 1972) and yet, despite significant population growth and industrialization over that period, oil and natural gas has continued to flow.

Recent debates around resource flows are taking place in a context of significant demographic shifts – including a growing global population, an ageing western population – increasing urbanization and the spread of urbanization processes to the newly developing economies of China and India in particular, deforestation, energy constraints, the loss of biodiversity, the impacts of climate change for global populations through flood risk, extremes of temperature and water shortages. Economic growth and the geographical spread of modernized and industrialized notions of economic activity and associated energy use to newly developing countries means that one estimate is that the world will use twice as much energy in 2035 as it did at the start of the twenty-first century and that by as soon as 2020 electricity demand could be 70 per cent higher than two decades previously (Roberts 2004).

Yet there is a fundamental unevenness at work here when 25 per cent of the global population lacks access to electricity or fossil fuels while a disproportionate amount of global energy is consumed by western nations (Roberts 2004). Levels of western domestic consumption of goods and services generally mean that in a fossil fuel-based economy significant levels of oil and gas are frequently imported. The terms of the debate on resource futures rarely question in any fundamental way the levels of consumer activities within western economies. Embedded in many western lifestyles is a view of (material) progress that is underpinned by the ubiquity of energy, rapid consumer-based growth and asymmetric producer–consumer relationships in relation to resources and in the production of carbon dioxide and greenhouse gas emissions.

The organization of resource flows is predicated on such a view of what constitutes progress and attempts to respond to the challenges posed by resource constraints and climate change need to be seen in light of the sociopolitical organization of the huge metalogistical systems (Luke 2003) that make fossil fuel economies possible. To take the example of oil and its relationship to western lifestyles, forms of growth and economic activity: this rests on complex relationships between oil wells, supertankers, pipelines, coalmines, power plants, markets, transmission lines, cars, trucks, trains, ships, gas, coal, heat, mobility, regulation, foreign policies and so on. There is, in short, a significant degree of lock-in to existing systems and ways of doing things and interrelationships globally means that shocks in one part of the system often have implications for other parts of systems (see Roberts 2004).

Within these wider global webs and interrelationships, a critical issue emerging for national governments is to ensure both secure and continued access to the resources needed to ensure their economic and social reproduction and also the 'right' type of resources to ensure a meeting of commitments to international agreements and targets, the long-term environmental commitments of a country and its ecological well-being. These new pressures encompass a multiplicity of issues which we can understand as follows. Security concerns in relation to infrastructure in the post-9/11 western world were primarily concerned with the prevention of physical, chemical and cyber attacks on infrastructure systems. The geopolitics of the contested but coming peaking of oil, natural gas, constraints on water supplies, the implications of the price of responding to climate change, of setting a carbon price and regulating activity in the pursuit of climate change-related aspirations redefines the parameters of security of infrastructure systems. It means that for national governments conceptions of national security may become broadened and extended. This not only includes defending infrastructure from foreign state or terrorist attack but also means that 'security' extends to security of supply of resources and securing the material, economic and social fabric of a country.

This raises the fundamental issue of whether these attempts represent broadly a continuation of fossil fuel-based economies and their associated growth regimes or whether a radical transition to a low carbon economy is envisaged. What role do technologies, technological fixes and behavioural change play? Are they technocratic responses to these pressures or aligned with fundamentally different philosophies as to the type of society or societies that should follow? Of course the issue here is not only one of national governments seeking to secure resources but also for what purpose, where, how, when and for whom. The issue of security in this sense is primarily one of power and its mechanisms, processes and effects. It is concerned with the circulation of production and consumption and how that is spatially organized.

> The specific space of security refers then to a series of possible events; it refers to the temporal and the uncertain which have to be inserted within a given space. The space in which a series of uncertain elements unfold is, I think, roughly what one can call the milieu . . . The apparatuses of security work, fabricate, organize, and plan a milieu even before the notion was formed and isolated. The milieu, then, will be that in which circulation is carried out. The milieu is a set of natural givens – rivers, marshes, hills – and a set of artificial givens – an agglomeration of individuals, of houses, etcetera. The milieu is a certain number of combined, overall effects bearing on all who live in it. It is an element in which a circular link is produced between effects and causes, since an effect from one point of view will be a cause from another.
>
> (Foucault 2008: 20–21)

Increasingly this creates pressures for an emerging set of relationships to be organized in respect of the trumpeted resource efficiencies of energy, water and waste systems that are organized in circular ways that bring production and consumption physically closer together thereby addressing transmission and distribution losses through geographically compressing infrastructure systems and also through behavioural change strategies. Yet this is a relative view of reconfiguration in that the very networked infrastructures that support global urbanism through the interconnections of places may also be being re-enforced through information and communications technologies, aviation networks and also through network-based trials of new technologies.

These new pressures are, of course, mediated through the power relationships of existing urban growth coalitions – organized across multiple levels and in particular the ways in which national state actors view places – and also through existing socio-technical systems. This is particularly important as it says much about latent capacities at an urban scale to organize the relationships between cities and infrastructure mediated through these new priorities. It also says much about the ways in which national state actors are able to discipline state spaces through cascading down regulatory measures and the extent to which urban coalitions are able to effectively respond to this or even prefigure such measures. These new pressures are appropriated and mobilized differentially by multilevel governance coalitions in different contexts. These pressures also encourage newer players to enter both urban governance and infrastructure system fields and, in some cases, working between and bridging the two. In this respect these new pressures are mediated through the path dependencies of existing systems and the transformative aspirations of both new players and entrepreneurial institutions and actors within existing coalitions.

We are at the point of ideological struggle here. We are faced with the challenge of governing the (re-)organization of production–consumption relationships; or, to put it another way, of resituating the city in a space of circulation and the way such a process is governed (by whom, promoting what values, with what effects and so on). Yet this is to be done an age of the decentring of power in which governments have contributed to their own ungovernability.

Representing, reproducing, repositioning space and place

What we seek to do here is to develop a framework for understanding the governance of space, of resource flows and socio-technical infrastructure systems and the dynamic and ongoing ways in which this shapes places and vice versa. Given the neoliberal imperative of a constant search for transformation and innovation and the re-emerged role of cities and regions within the latest phase of globalization, how can we understand the ways in which these new and emergent pressures impact and are acted upon at an urban scale? That is how it is that these new pressures are appropriated and responded to by a variety of multilevel governance institutions and actors – national state actors and institutions, (elements of) urban coalitions, representatives of socio-technical infrastructure systems and how these social interests are organized, how they envisage future relationships between resource flows, their production and consumption, particular places and the socio-technical infrastructure systems that mediate such relationships.

The tensions between what things are and what they can become is pronounced in the face of new ecological pressures – climate change and resource constraint – and the ways in which these are incorporated into the relational spaces that reproduce the relationships between cities, infrastructures and resource flows. Yet, if responses to these pressures are not natural, who are the social interests involved in promoting the status quo or transformation? If our understanding of relationships between urban contexts, infrastructures and resource flows is not reduced to 'endogenous' and non-exploitative relations, how do we understand responses to new pressures? And, given that – in a context of urban entrepreneurialism – there is a constant search for transformation and innovation, how do cities seek to present this transformation, to whom is this communicated and with what consequences?

These three questions underpin a concern with not only understanding processes of transformation but also addressing constraints on and urban contexts of transformation. We raised the issue earlier in this chapter about the ways in which new pressures are likely to bring responses from coalitions that includes both established actors within urban and infrastructural fields and also new actors and social interests. We are thinking here of the multiplicity of relationships that not only constitute the city, but also produce socio-technical energy, water, waste and transportation systems – which in a context of privatization and liberalization can be best understood as encompassing a range of social interests and actors. The development of urban strategies is thus likely to include well established interests in the fields of both urban governance and the governance of socio-technical systems and new entrepreneurial actors working between the two. A relational understanding opens up space as political. In the process of becoming and constituting new identities, spaces, places and reimagining the communities and economies of cities (Anderson 1991; Cameron and Palan 2004), '[r]ather than an explicit conflict over political aims what we have now is a confrontation between imaginations of the city' (Massey 2005: 157).

In the process of becoming, the opening up of relational space in the political reproduction of space and place begs the critical question: how open? This is a critical issue because it fundamentally addresses questions of who can and does participate, how and on what basis are these different interests and their expectations negotiated? These questions can be mobilized analytically to address the dominant representations of places, the ways in which it is envisaged that resource flows and socio-technical infrastructures are reconfigured, where and to whose benefit and detriment, and how this representational space is produced. At an analytical level this allows us to understand representational space through bringing it together with relational space and relative space.

The power of such representations are in their search for certainty, abstraction and universalism in the relationships between cities, socio-technical infrastructures and resource flows and attempts to achieve (often temporary) closure or stabilization (Bijker et al. 1987). There are, thus, two intricately linked issues here. The first relates to trying to gain greater understanding of the processes of constructing representations. The second is in trying to understand the ways in which such representations frame a partial, privileged understanding of relationships between cities, socio-technical infrastructures and resource flows.

The process of framing (Goffman 1974) representations is a collective practice which involves entangled webs of human relations, institutions, artefacts and so on. Framing is concerned with establishing 'a boundary within which interactions – the significance and content of which are self-evident to the protagonists – take place more or less independently of their surrounding context' (Callon 1998b: 249). This requires the drawing of boundaries 'between the relations which the agents will take into account and which will serve in their calculations and those which will be thrown out of the calculation as such' (Callon 1998a: 16). Framing allows for the definition of individuals, groups, objects, goods and so on in that they can be disentangled or disassociated from entangled webs and relationships.

Through delineation, framing 'puts the outside world in brackets, as it were, but does not actually abolish all links with it' (Callon 1998b: 249). Yet, this should not be taken for granted in that 'far from limiting the possibility for political conflict and negotiation, framing forms something like a surface on which forms of political reflection, negotiation and conflict can condense' (Barry and Slater 2002: 185). This is a concern with the politics of closing down, of seeking to persuade through fixing a view of what the city and its relationships with socio-technical infrastructures and resource flows are. An opening up of relational space is essential to understanding what these relationships can become.

What they are and what they can become is related to the issue of the relationships, knowledge and financial capacity that can be mobilized into a capability to act. The tension here is between closing down and fixing views of place through privileging representational space and obscuring the relational and relative spaces of its production and the opening up of relational space to rethink relative and representational space. At the extremes it is a tension between the status quo, its obduracy and transformation, between narrowly constituted private space and a more broadly conceived participatory public space and between inclusion and exclusion. It is useful here to invoke the idea of relational and relative space as fields (Bourdieu 1993).

Field helps us to think about relational space as the capacity to act (that is latent action or possibility) and the capability to act (which is the mobilization of capacity). Thinking in this ways helps us to understand the lattice of governance arrangements, the multiplicity of different actors, public and private, at different scales, supranational, national, city-regional, organizational, with often differing motivations, forms of knowledge, financial resources and aspirations who take a view of, make claims of and act on the relationship between city, infrastructure and resource flows. This we see as 'a field of multiple actors, trajectories, stories with their own energies – which may mingle in harmony, collide, even annihilate each other' (Massey 2007: 22).

Institutions are constituted by social relations of regularity, routines and norms and through which social action is often performed unreflexively. Yet, what was historically appropriate in one era may become inappropriate in a context of transformation. In the early stage of attempts to defend or transform relationships between cities, infrastructure and resource flows the degree of institutional lock-in among key policy agenda setting institutions and organizations is important, including the role of national government departments, key agencies, business, local authorities and industry. With this in mind, it is important to understand the role of transformative agency, in relation to particular institutional settings that are often constrained by institutional lock-in, in (re-)producing cities, infrastructure and resource flows.

This requires thinking about institutional interrelationships of knowledge, financial resource and capabilities in relation to and as a response to new pressures and, in this respect, the entrance of other influences (often national state actors) into the field of relationships that constitute urban infrastructures and resource flows. Understanding the responses to new pressures requires an acknowledgement of interrelationships across scales of political activity – both as historically constituted and how they might be reconfigured. This needs to take account of the institutional context of those who become involved in these processes but also how from these positions they seek to build relationships. These relationships have consequences for the forms of knowledge involved in attempts at reproduction of radical transformation, the financial resources that can be leveraged in these processes and symbolically what this says about the city to others.

Spaces of reproduction and radical reconfiguration

The overriding issue that came out of the previous section was the centrality of thinking about (re-)producing or transforming cities, infrastructures and resource flows in terms of institutional interrelationships. While there has been an acceptance that 'institutions matter' (Jessop 2001), there has, however, been a lack of clarity in defining institutions and a wide variation in how institutions are defined but also in the respects that they are held to matter and the reasons for suggesting they do (Jessop 2001: 1213).

Institutions are not *necessarily* synonymous with organizations but, following Pierre Bourdieu (1991), should be seen as 'any relatively durable set of social relations which *endows* individuals with power, status and resources of various kinds' (Thompson 1991: 8, original emphasis). Institutions, through ongoing processes of reproduction and the struggles of a variety of actors in constituting them, never reach closure and are always provisional. While notions such as durability in thinking about institutions may be fruitful in relatively stable contexts of reproduction, with a resultant degree of institutional lock-in, how may we understand institutional interrelationships where transformation and innovation are being encouraged?

In contrast to structurally informed durability in social relations (the position outlined above) we can think about relationships between cities, infrastructures and resources as socio-technical (Hughes 1987). Existing institutional frameworks that produce cities – by definition – will often be predicated on an institutional reproduction that is relatively habitual. In the face of new pressures this can create tensions between the obduracy of existing institutional interrelationships and the possibilities of socio-technical transformation. This requires bringing exploring the intersections of institutional durability and socio-technical agency.

The emphasis on the durability of social relations highlights the importance of historically generated relationships, practices and knowledge and in many ways equates with what has been called lock-in. It also suggests that if we wish to understand the transformative potential of building new institutional interrelationships we need to take account of the ways in which those who become involved in responding to the new pressures, from particular institutional settings, are disposed to act in particular ways which have been informed by those settings. This is as dispositions are informed by social structures (Bourdieu 2000); social structures which are relational in the sense that they are bound up with the field of institutional positions (Bourdieu 1993).

But in the search for permanent innovation the danger is that historically informed institutional interrelationships, which may have been cultivated and appropriate for an earlier period of urban growth and/or infrastructural development may offer inappropriate social structures for those engaged in contemporary responses to new pressures. An issue is to what extent those engaged in socio-technical innovation are able to reflect upon these social structures, assess the possibilities for and limits upon transformation and then the shape subsequent action takes.

The historically structured relationships that produce a categorization of the relationship between city, infrastructures and resource flows may at the extreme of the habitual reproduce themselves – what we can call closed spaces of reproduction – while at the other extreme they may be radically transformed through the interpretation and reflection of the latent capacity that constitutes the field to form a very different capability to act – this we may see as open spaces of radical reconfiguration. The possibility for individuals and institutions – key players, consultants, etc. – to position themselves relates to the types and stock of relevant capital they hold.

As has been highlighted an important way of understanding the 'inside' and the 'outside' – of inclusion and exclusion – of the relationship between city, infrastructure and resource flows is through the notion of field. Fields can be seen as 'structured spaces of positions (or posts) whose properties depend on their positions within these spaces and which can be analysed independently of the characteristics of their occupants (which are partly determined by them)' (Bourdieu 1993: 72). In addition to 'general laws' of fields, where we can 'use what we learn about the functioning of each particular field to question and interpret other fields' (Bourdieu 1993: 72), there is a particularity to every field. Pertinent here is that where established interests may be confronted with new players,

> we shall find a struggle, the specific forms of which have to be looked for each time, between the newcomer who tries to break through the entry barrier and the dominant agent who will try to defend the monopoly and keep out competition.
>
> (Bourdieu 1993: 72)

These struggles are informed by and informing of power relations within the field. In the course of struggles players accumulate various forms of capital 'which orients subsequent strategies' (Bourdieu 1993: 73).

The extent of either new actors in the field or those who are able to 'step outside' of their institutional setting is crucial. From this we can understand the balance between institutionally conditioned and reflexively negotiated interrelationships that underpin responses to these new pressures through four interrelated concepts, which take account of not only social relations involved but also the knowledge, skills and economic capital that are brought into developing responses and which are subsequently cultivated (Bourdieu 1986).

The first of these, social capital, allows us to focus on the spatial development of social networks which are not pre-given but the product of 'investment strategies, individual or collective, consciously or unconsciously aimed at establishing or reproducing social relationships that are directly usable in the short or long term' (Bourdieu 1986: 249). The second, cultural capital, allows us not only to think about these social networks spatially, historically and through investment strategies but also to explore what forms of knowledge – e.g. technical, policy, lay, local – and skills are possessed and developed within these networks.

Cultural capital may exist in three forms (Bourdieu 1986: 241–258): the embodied state (in dispositions of officials and policy actors), the objectified state (in cultural goods such as promotional literature, policy statements, business strategy documents, and also machinery and technology) and the institutionalized state (for example, carbon reduction frameworks). The third concept, economic capital, is that which is convertible to money, property and so on and permits us to think of a further facet of the structuring of urban social relations. The tension between the obduracy and transformation of urban contexts may be understood through the interweaved negotiation of attempts to cultivate social relations, forms of knowledge and skills and economic capital which both underpin views of urban infrastructural transformation but which are situated within wider institutional contexts. This, then, informs the fourth concept, symbolic capital, which refers to the ability to confer meaning or *legitimization* on a particular process of urban infrastructural (re-)production.

The process of becoming through the relational space of the field is thus one of linking the politics of participation to the production of meaning of the relationships between city, infrastructures and resource flows. Yet within the production of the relational space and the actor-networks that constitute them is also a view of the relation of science to politics and of expertise to democratic participation and the roles in decision-making of technocratic expertise and democratic participation. This begs the question: how should this be decided? And relatedly, how far should participation go and how should modes of expert knowledge be integrated with local forms of knowledge? This creates the possibility for bringing together in relational space a concern with technical possibilities and political priorities in the pursuit of what ought to be done.

Bringing together technical forms of knowledge and localized cultural knowledge requires the work of mediation or brokering between different and often disparate social interests from institutional contexts. The act of performing this role of mediating communication between these different social groups requires those performing the role to have enough information and understanding to be able to explain the concepts of one community to another, including the meanings assigned to particular technical objects and artefacts and being able to communicate in the different languages groups use (see Fischer 2003). In creating relational space in this dynamic way, the boundaries that are often fixed and closed in representational space have the potential to be opened up. It is in this respect that institutional structure and transformative agency come together. The potential for obduracy is when those who may be involved in developing urban infrastructural innovations are unable or unwilling to step outside of their institutional context and become cultural dupes to the social structures which inform their dispositions to act (Bohman 1999).

The ability to reposition themselves, for those officials, policymakers, industrialists and so on who become involved in urban infrastructural innovation, suggests some (differential) capability to act in more innovative ways within existing cultural institutional constraints. This more reflexive move allows actors 'within' an urban context the *possibility* to reflect on cultural constraints and to seek where appropriate to innovatively reposition themselves in terms of the new relationships that constitute a response to emerging new pressures. In this view urban infrastructural innovation is culturally negotiated and open to interpretative flexibility (Bijker et al. 1987). The issue is how certain configurations win out in the vast array of innovative possibilities or how heterogeneous urban infrastructure networks come to be assembled (Law 1992).

Representing and communicating urban infrastructure and resource flows

It is important to consider the embodiment symbolically of these negotiations and how these are presented and communicated to others and other – relative – spaces (one may think particularly of how urban infrastructural innovation is communicated to external actors including civil servants and policymakers). Importantly, relational spaces inform both the spheres of production and reception of symbolic exchange (Thompson 1995). This is pertinent where those closely involved in the relational production of an agenda for urban infrastructural transformation, seek to communicate representations of this to external others, such as national level civil servants, competitor cities and funders. This may be done through the embodiment of agendas and visions in promotional literature, conference presentations, through media reports and website materials (embodied cultural capital). It may also be communicated at a face-to-face level or via telephone and other mediated communication.

Relational spaces of reception may also have implications for the receipt of such representations which may influence the relational production of the space of reception. This allows us to think not only how new pressures are produced and communicated but also how they are received by those claiming to speak on behalf of other cities and how this affects relationships with socio-technical infrastructures and resource flows. In terms of spaces of reception, the institutional context and social attributes of those receiving – that is the acquired skills, dispositions and the resources available to social interests – 'are key elements in a process of appropriation that begins with the initial reception of media messages but extends well beyond it, involving other contexts, other individuals, other messages interwoven with those initially received' (Thompson 1995: 111). Tensions of obduracy and transformation exist not only in the production of relational space and place but also in how this is received by others. This brings questions of agency and transformation together with the structured relationships within and between institutions, and is important to looking at the opportunities for and constraints upon urban infrastructural innovation.

Conclusion

In this chapter we have examined the basis of relationships between cities, infrastructures and resource flows. We have demonstrated that they are politically produced through spatially configured coalitions of interrelationships of institutions and social interests that are dynamic and always in the process of becoming even if that process is punctuated by temporary fixes. In the post-war Keynesian-Fordist period in countries of the industrialized world, many of these coalitions and the institutions and social relations constitutive of them were relatively settled and were organized around equitable principles such as universal service and a broad understanding of the national interest.

The ideological challenge provided from the 1970s by neoliberalism questioned the relative certainties of the administrative provision, the organization of comparatively plentiful natural resources and the strategic view of spatial equalization inherent within it. The promotion of uncontrollable neoliberal processes of globalization was, paradoxically, ideologically and politically driven. A rolling back of the Keynesian-Fordist nation-state met with a rolling forward of a quite different form of state predicated on entrepreneurialism, innovation, privatization, liberalization, deregulation and economic competitiveness.

As part of this institutional struggle and reconfiguration, the relatively benign role of both cities and infrastructure were opened up and, particularly from the 1990s onwards, there was a re-emergence of the urban and regional scales 'as key institutional arenas for a wide range of policy experiments and political strategies' (Brenner and Theodore 2002: 341). In particular notions of urban entrepreneurialism and competitiveness were promoted as a means of encouraging innovation and the remaking of places. In doing this the global and supranational institutional frameworks and regimes of accumulation that had been subjected to neoliberal ideological assault, reorganization and cultural transformation provided the wider conditions within which urban strategies were formulated. These wider institutional shifts created the parameters within which urban strategies are conceived.

Critically important here are, as we demonstrated, relationships between national states and territory. In a neoliberal context, at its extremes, it sees the state privileging particular relationships that are viewed as exemplary and motors or engines of national economies and alternatively those that are increasingly marginalized places where the role of the national state is to make places amenable to inward investment. There are numerous positions between these extremes that allows the relationship between national state and subnational territories to be characterized in terms of a variety of new state spaces (Brenner 2004).

Responding to the pressures to be entrepreneurial and competitive at an urban level can be understood differentially not only in terms of the latent capacities that can be mobilized locally and innovatively but also in respect of the ways in which different places are viewed by national governments, supranational political institutions and multinational capital. Cities, thus, find themselves in ongoing processes of permanent innovation the purpose of which is often less than clear other than the reproduction of neoliberal tenets of economic growth through competition, entrepreneurialism and innovation. In practice this means the constant reinvention of institutions, governance arrangements and latent local capacities to act entrepreneurially and innovatively in anticipating to and/or responding to changing national, supranational and multinational capital requirements and regulatory frameworks. It also means that the infrastructural underpinnings and the organization of resource flows that inform the constant search for urban entrepreneurialism are also themselves subjected to that very urban entrepreneurialism. Post-Keynesian infrastructures, in western, industrialized contexts have become variable, competitive and splintered from the Keynesian-Fordist ideal of universal service knitting national space (Graham and Marvin 2001). They both inform variable economic growth and competition and are subjected to its principles.

With this background of wider conditions and landscape pressures in mind, neoliberal urban strategies are increasingly confronted by a new set of pressures, where the resources and modes of infrastructural organization that underpin spatially variable neoliberalizations or neoliberal manifestations have become subject to challenge. The forms of urban economic growth that the organization of resource flows and infrastructure have mediated are subject to significant ecological challenges: through the pressures posed by peak oil and the wider geopolitics of oil and natural gas, water constraint and stress, the environmental consequences of landfill waste, the carbon and methane emissions associated with contemporary modes of economic organization and the consequences that this poses for urban areas.

It is our view that the organization of relationships between cities, infrastructures and resource flows, in addition to being organized in ways which seek to secure economic growth, will increasingly be mediated through pressures of climate change, resource constraint and attempts to secure access to natural resources. Not only will this be the case but also the hitherto mentioned variability between places and where they are located within a multilevel governance maelstrom will mediate these pressures through which urban strategies will be negotiated and formulated. This is a question of politics that takes places not in a context where history is erased – however much that often appears to be the case – but where existing power relationships mediated through institutions, infrastructures and resources condition but do not determine the shape of urban responses. To put it another way, these pressures to reconfigure relationships between cities, infrastructure and resources take place in relation to existing spaces and places where obduracy and transformation structure local struggles.

This matters in that it brings to the forefront of this debate the issue of whether responses are formulated through a status quo of social interests or new social interests and which of these – or new combinations between them – is preferable. That is to say that forms of urban infrastructures and resource flows are mediated not only through neoliberal forms of economic activity but also through climate change, ecologically extended notions of security, resource constraint, spatial regulation, national state priorities and so on but these discourses may be appropriated and are mobilized differentially by coalitions in different spaces.

This will be due not only to the relative power of social interests to impose their expectations of what urban infrastructures and the resources that flow through them are for but also in certain urban contexts the lack of power to mobilize alternative expectations. Not only is the issue one of the power to define urban infrastructure and resource futures in relation to particular cities but also the power of particular urban coalitions to produce a view of a city's relationship to infrastructure that positions it more favourably relative to other cities. In this respect the competitive race of urban entrepreneurialism is being opened up to a new struggle as to what the urban eco-entrepreneurial race would look like. Inherent in such struggles are the ways in which cities and socio-technical infrastructures are organized and governed and views of and value judgements of the ways in which relationships between society and technology, urban and nature, economy and ecology, and scales of governance are mediated through this and remade.

The issue here is one of governance, values and power in seeking to represent future views of the relationship between cities, infrastructure and resource flows, their translation and the consequences of this. This, as we demonstrated in this chapter, requires ways not only of understanding these new pressures but also in thinking about the relationships between them and strategic urban responses that seek to imagine a new urban future from the positions of existing relationships and contexts; that is the connection of a strategic response of a different urban future from existing and historically produced contexts and requires addressing the relationships between obdurate urbanisms, the potential for transformation and the dynamism inherent between these two broad positions. In short we are talking about the ways in which we can rethink our understanding of the spatial organization and governance of cities and infrastructure in a context of these new pressures. It is to an examination of these new pressures that we now turn.

2

'New' pressures:

Mediating cities, infrastructures and resource flows

Introduction

Drawing on both academic and policy literatures, this chapter reviews the ways in which anthropogenic global environmental change is creating a set of pressures that are mediated through two dominant urban processes. The first of these we characterize as a national security discourse. Critical infrastructure protection and urban resilience debates, we argue, are being extended from a concern with the military security of a particular infrastructure, the vulnerability of an infrastructure to terrorist attack, to wider concerns about the reliability and security of the energy, water, food and waste resources that pass through the entire system. This signifies a fundamental concern with how 'secure' whole infrastructure systems are under the changed conditions of climate change and resource constraint.

The second issue we address relates to how ecological questions are mediated through national frameworks of urban and regional governance that are mainly concerned, as we discussed in Chapter 1, with ensuring national economic competitiveness and innovation. Here we find a different set of concerns focused on understanding how climate change and resource constraint affect the abilities of cities to secure economic and material reproduction. Given the territorial unevenness inherent in the raison d'être of the competition state, when national state institutions and actors are confronted with the challenges posed by climate change and resource constraint how this then affects relations with the national space economy – in particular the differences between more privileged and less valued spaces – is a critical issue. This extends not only to these relationships but also beyond, to the ways in which climate change and resource constraint affect the continued replication of key sectors and linkages within and between cities. This additionally has implications for the ways in which and the extent to which, in a context of permanent innovation, these pressures offer opportunities for new rounds of innovation and reproduction. Taking these two processes together as mediating the ecological effects of climate change and resource constraint we argue that these then create significant new pressures for future relationships between city, infrastructure and resource flows.

Our argument unfolds as follows. We explore a series of urban threats to city, infrastructure, resource flow relationships that both contribute to the production and consequences of climate change impacts. The chapter then moves on to address how two dominant logics – an extension of a national and military security discourse and a state national economic competitiveness discourse – create pressures for strategic responses at an urban scale that seek to try and guarantee resource security as fundamental to continued economic and material reproduction. We finish the chapter with key conclusions.

Global climate change 'impacts' on cities and infrastructure

We now examine how the global ecological changes significantly produced by global urbanism create pressures that are likely to reshape cities and their relationships to infrastructures. In particular we are interested in how global ecological change seems to be reshaping the very context within which the economic, social and material reproduction of cities are to be secured. Although urban implications are relatively poorly understood in most places, the very largest world cities are trying to develop new forms of knowledge and understanding of the longer-term implications of ecological change on climate and infrastructure.

Climate change science is now unequivocal that the warming of climate systems is taking place and that, unless there are significant reductions in the global emissions of greenhouse gases, wide-ranging climate change impacts such as sea level rise and more unpredictable weather patterns will follow (Intergovernmental Panel on Climate Change (IPCC) 2007). These changes are likely to lead to significant environmental, social and economic impacts across different sectors and places that include effects on agriculture and fisheries, desertification, biodiversity, water resources, energy supply, heat and cold related mortality, coastal zones and flooding. At the same time there is critical recognition that these impacts are most likely to be felt in major cities and urban areas. Cities are key centres of economic activity and given their potential responsibility in the production of carbon emissions, estimated at up to 78 per cent (Stern 2006), they are a key site through which to consider the impact of climate change (see the review by the Organization for Economic Cooperation and Development (OECD) 2007).

Rescaling climate change and cities

Climate change is likely to have variable implications for different cities. For example, the Intergovernmental Panel on Climate Change's third report (IPCC 2001) concluded that climate change was most likely to have important impacts on cities in coastal and riverine locations and resource dependent regions. The IPCC's (2007) fourth panel extended this to cover locations at risk from extreme weather events, especially those where rapid urbanization is occurring and those areas whose economies were dependent on climate sensitive resources such as energy, food and water. Critically then climate change could affect the continued material and economic reproduction of cities if it increases the vulnerability of key resource flows. There are however serious difficulties in understanding how climate-change-related pressures will inform the reshaping of cities and regions.

The development of serious long-term understanding of such processes is problematic, not least as conventional global and regional climate change models are not able to easily deal with cities. Conventional models are complex mathematical simulations of the atmosphere, oceans and land surfaces through exchanges of heat, flows of water and energy and the primary tool for predicting the response of climate to increased greenhouse gases. Although cities contain half the world's population, urban regions cover only 0.2 per cent of the earth's land surface and the diversity of cities' land surfaces are not adequately dealt with by conventional climate change models (Jin and Shepherd 2005). Consequently, developing accurate and reliable understanding of the implications of climate change is highly complex with existing global and regional models of climate change and considerable effort is being expended into trying to develop more scale-sensitive models that can model and forecast impacts at urban scales and below (Jin and Shepherd 2005).

Yet, despite this there have been a number of efforts to undertake more detailed analyses of city-scale impacts of climate change across different sectors, in most 'notably London, New York and Boston' (OECD 2007: 6). These studies included detailed estimates of potential impacts, undertook quantitative and economic analysis and gave consideration to different adaptation options. There are also 'impact' studies in other developed cities including Toronto, Wellington, Sydney and Melbourne, limited city-scale analysis as part of wider regional studies in Los Angeles and major cities in Europe, and a large number of cities with partial and/or qualitative analyses. While there has been wide coverage across different continents (see Figure 2.1) most studies have focused on coastal cities, but there is comparatively less information available on inland cities and some key coastal regions have poor coverage. There is 'a major evidence gap on the impacts of climate change across the range of geographical locations and impact categories, such as energy demand and water resources' (OECD 2007: 6). Consequently, although the research effort is underway the current literature is only indicative of the issues that are expected to shape cities.

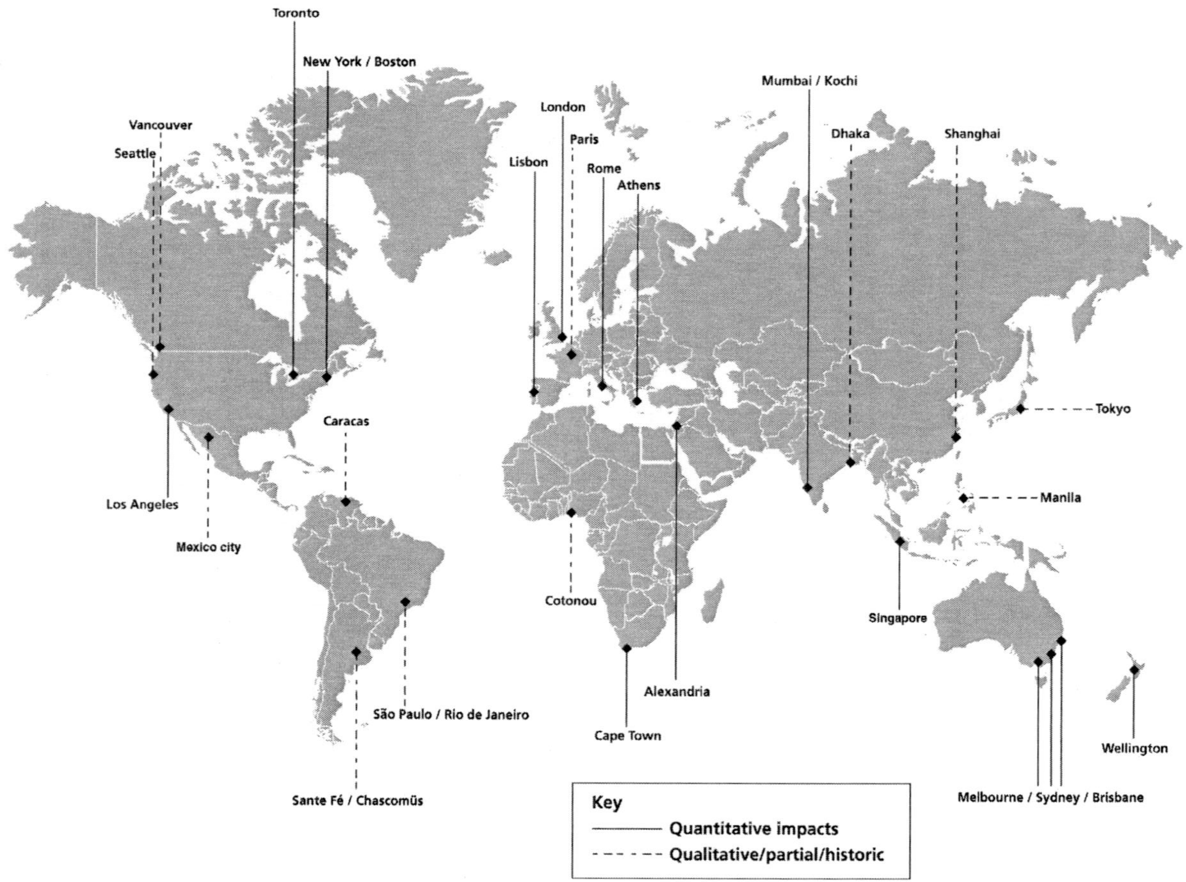

Figure 2.1 Major city climate change studies.
Source: OECD.

Despite these problems, existing analyses show that it is not just coastal–riverine locations and dependent cities that will be most affected. There are also expected to be serious challenges for inland cities. At the same time, in addition to extreme events there are likely to be significant implications related to urban food, energy and water use. The economic effects are uncertain but potentially very significant and there are considerable differences in the expected impact in developed and less developed megacities (Bicknell et al. 2009).

Climate risk in cities

Here we are primarily concerned with 'first order' impacts and how climate change directly affects the wider environmental context within which the city sits. This largely involves the direct effects of sea level rise on coastal cities, built environment and infrastructure damage from extreme weather, effects on energy use, including heating and cooling energy for water, water availability and resources, food and water and vector borne disease. There are of course another set of important effects that climate change may have on the availability of physical assets used within cities, in particular the costs of and reliability of materials and resource inputs into economic production. This may then affect costs to businesses and the competitiveness or the comparative advantage of different cities. These issues are reviewed in the section below on carbon control.

Over 50 per cent of the world's population and major world cities are located in low lying areas or near coasts and so are vulnerable to sea level rise and storm surges. By 2100 global mean sea level rise is expected to be between 18 and 59 cm but could be higher under later scenarios (Nicholls 2004). There is much greater certainty attached to the probability of sea level rise than other climate change impacts and, as such, the issues have been better researched. Coastal cities frequently contain large human populations and are key national and international economic centres (McGranahan et al. 2007). A number of these cities are already below normal high tides levels and prone to flooding. Climate change is likely to intensify the problems for these cities through sea level rise and the frequency of extreme weather events. This will cause particular problems for cities in deltas, low coastal plains, islands and estuaries including rising water tables, increased storm flooding, inundation and poor drainage.

Urban infrastructure and the built environment of cities are vulnerable to the impacts of extreme events, primarily floods and storms, and also including heatwaves and drought. These are likely to be most problematic in the largest and low-lying coastal cities. Storms are extremely costly and climate change is likely to result in significant increases in potential future costs. These have been estimated by 2080 to result in a 65 per cent and 75 per cent increase in costs during severe hurricane seasons in Japan and the United States. Many estimates of the cumulative contribution of enhanced climate change risk point to a doubling of worldwide economic losses due

Figure 2.2 The effects of climate change are laid bare in an online simulation which shows gamers the impact of flooding on London.
Source: Floodsim.com

to 'natural disaster' every ten years (OECD 2007). While global assessments of coastal flooding and mean sea level rise are well advanced, there have been few detailed studies with a city focus that have attempted to undertake local modelling of the effects of sea level rise on the city (see Figure 2.2). A key exception to this is work undertaken in London to assess the impacts of sea level rise on the River Thames up to the end of the twenty-first century (Thames Estuary 2100, 2007).

Although there are few predictions of storm losses, especially at an urban level, work on the implications of Hurricane Katrina and New Orleans has developed a better understanding of the implications of such storm events on cities. There are numerous estimates of the costs of the damage of Katrina, ranging from $100 billion to double this figure once the additional effects of oil price rises, increased energy costs and other factors are taken into account (Burton and Hicks 2005). Additionally there is serious concern that climate change will increase the scale and frequency of intensive rainfall events and related flood damage with other factors likely to increase flood damage, including rising population and the urbanization of flood plains. There have been a number of attempts to approximate the costs of urban flooding. For example in Boston these were estimated by 2100 to exceed $57 billion, based on an assumption that no adaptation was undertaken, of which about $26 billion was attributed to climate change. It was claimed that a proactive adaptation strategy could reduce the cost attributed to climate change to $9 billion (OECD 2007). Overall the cost of making the built environment and infrastructure resilient to all climate change 'impacts' in

OECD countries ranges from between 0.005 per cent to 0.5 per cent of GDP ($15 billion to $150 billion) per annum with high costs possible if there are higher temperatures. OECD countries currently invest approximately 20 per cent of GDP in infrastructure and the Stern (2006) report estimated that additional spending of between 1 and 10 per cent was required to limit damage from climate change.

Changing weather conditions are expected to result in increased winter heating and summer cooling. The implications of this for Europe and the United States are thought to be modest in the short to medium term. However, at a finer scale the effects are likely to vary over space. So, for instance, it is anticipated that there will be a strong increase in demand for cooling in southern Europe while reduced energy demand for winter heating will occur in northern Europe. Even more pronounced implications at the city scale may be anticipated because of cities' disproportionate concentration of energy demand and additional factors such as the urban heat island effect that can potentially increase demand for summer cooling. Already city authorities in major world cities are developing heat maps of the city to better understand and manage these implications, such as those in Tokyo and London.

A number of approximations have already been made of the likely changes in energy demand. For instance in London it is estimated that energy use for cooling in an air-conditioned office will increase by 10 per cent by the 2050s and 20 per cent by 2080. In Boston it is expected that by 2030 climate change will be responsible for 25–40 per cent of increased energy demand in the region and in California it has been predicted that extreme heat events will increase rapidly implying that heatwave conditions may last for the entire summer by 2100 (OECD 2007). Projected increases in peak demands for electricity will be difficult to meet because of potential shortfalls in transmission and supply capacity. The effects associated with summer temperature rise will be more important in cities in hot climates where 60 per cent of total electricity use in the commercial sector may be responsible for cooling (IPCC 2001). There are also likely to be issues of increased energy use for water supply, particularly in respect of pumping, desalination and water transfers in some regions. Consequently there are complex effects on cities that are currently not well understood. Even relatively small increases in energy demand are likely to result in substantial extra expenditure on energy that might have other knock-on effects on city economies and competitiveness.

Climate change is likely to affect the demand for, as well as the availability of, water resources. Increasing temperature will accelerate the rate of evaporation and both the demand for cooling water in cities and increase demand for water. This will particularly affect semi-arid and arid areas such as the Mediterranean basin and the western United States, as well as developing regions like southern Africa. Climate change that reduces rainfall is likely to be a serious issue for cities in these regions, especially those with existing water scarcity and/or those dependent on snowpack and glaciers. There has been little detailed work about how change in water demand will affect cities and economic sectors. Where work has been done, studies have predicted that for the year 2030 annual water use in Washington DC is likely to increase by up to 11 per cent due to climate change (OECD 2007). There are, however, uncertainties about the impacts of climate change on levels of rainfall and drought requiring downscaling of models and forecasts. There is now significant interest, in cities that are likely to be affected, in developing adaptation strategies and debates about water autonomy.

Critically, then, climate change creates a more uncertain and difficult to predict context within which cities seek to ensure their continued reproduction. In particular it is the larger world cities in the global north that have been able to make the most significant investment in developing the knowledge and expertise to localize global and regional climate change models to assess their metropolitan implications. At the same time, however, there are also critical gaps in understanding how climate change affects metropolitan demand for critical resources such as water, food and energy consumption within the city. Additionally there are also uncertainties about how climate change affects the reliability of wider metalogistical systems of food, energy, water and materials that support cities. In the next two sections we examine how these pressures are reinterpreted and recast through a security and economic framework.

Urban security and Critical Infrastructure Protection

Critical Infrastructure Protection (CIP) is a concept that relates to national states' preparedness to develop a strategic and effective response to serious incidents that involve the critical infrastructure of a city, region or nation. In the United States CIP is a national programme that attempts to ensure the security of vulnerable and interconnected infrastructures and was established by the Clinton administration in 1998. Following 9/11 President Bush broadened the definition of CIP as the systems that are 'so vital to the US that the incapacity or destruction of such systems would have a debilitating impact upon security, national economic security, national public health or safety' (DHS Presidential Directive HSPD-7, December 2003). More widely, CIP programmes were developed at the European level and by other national states, most notably the UK and Australia (Little 2002). As initially conceived the focus was tightly drawn on understanding the vulnerability of virtual and physical infrastructure to terrorist or military attack.

CIP usually develops standardized definitions of critical infrastructure that includes road, rail, air and water transportation (including Just-in-Time systems, critical transport hubs and hazardous materials); power systems, including electricity, oil, gas and nuclear; information and communications technologies and networks; water systems and drainage; agriculture and food; as well as emergency and public services. Traditionally these infrastructures have been under public control, or at least public regulation, and have been largely separate and not highly interconnected. A series of

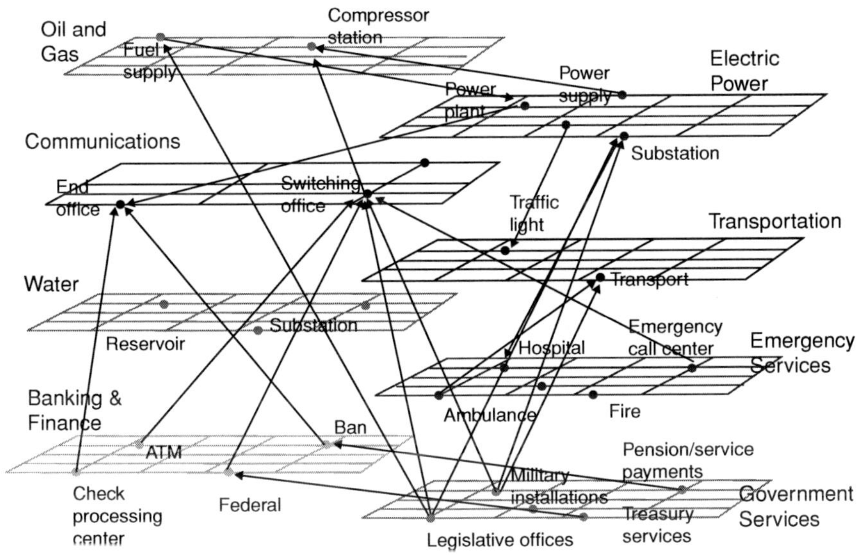

Figure 2.3 Interconnected infrastructure networks.
Source: Federal Communications Commission.

crises, however, illustrated the interconnected and vulnerable character of contemporary national infrastructures, including 9/11, the Canadian ice storm, and so on.

CIP was conceived to address increasing concern at national level and among governments that infrastructure systems were becoming more vulnerable to traditional and non-traditional threats ranging from technical failure, human error, weather and other 'natural causes' as well as physical and cyber attacks. In particular 'terrorist attacks' on infrastructure became a critical issue as infrastructure became seen as an easier target than conventional military targets because there was a much lower probability of detection. There was also clear evidence that infrastructure networks were much more interconnected than had often been acknowledged (see Figure 2.3). All infrastructures were reliant for safe and effective operation on energy and in particular information and communications technology (ICT) control systems. Such complex, and not always well understood, interdependencies meant that a catastrophic failure in ICT systems, for example as a result of physical and or virtual attack, could then have disastrous consequences in air traffic controlling, banking and financial, and electrical power systems, affecting the performance of national economies.

Furthermore, following key periods of privatization, liberalization and deregulation much critical infrastructure was new and owned and controlled by the private sector. Consequently, national governments potentially had relatively low levels of control over the 'preparedness' of infrastructure to attack or the coordination of responses in the event of breakdown and failure. Attempts to systemically increase resilience would require working across a wide range of commercial and public agencies at multiple levels. Yet there was often recognition that there was no unified national capacity for either understanding the vulnerabilities or increasing the resilience of the national CIP in any systemic or strategic way.

In the United States the Clinton directive led to the formation of a national structure for CIP whose primary action was the production of the National Infrastructure Protection Plan (NIPP). For each critical infrastructure sector the federal government appointed a liaison official and a private sector counterpart to develop a sector NIPP. In addition grants were made available from the Department of Homeland Security for municipal and private organizations to develop CIP – for emergency management, water security, rail, transit and port security, metropolitan medical responses, terrorism prevention programmes and the Urban Areas Security Initiative (UASI). The logic was one of protecting critical infrastructure from attack.

Urban vulnerability and rescaling infrastructure security

What is of particular interest for us is how large – especially national and internationally oriented – cities *and* infrastructural connections (airlines, freight, and telecommunications) have become a focus for critical infrastructure protection policies and strategies. While cities have always been a focus for security issues (see Graham 2004; Coaffee 2008) concerned with the protection of key business or governmental assets through cordons, control and military and political security there had not traditionally been the same focus on urban infrastructure networks or technologies. Although the initial focus of CIP was on the national scale, it subsequently became a debate that rescaled these concerns about infrastructure downwards on to the densest agglomerations of infrastructure, large cities, and then upwards to the infrastructure networks that ensure high levels of connectivity between cities.

This was for numerous reasons not least of which was an increasing recognition of the role of and interrelationships between cities and their infrastructure networks. The city was seen as providing an easy context for terrorists who could easily blend into the urban area and attack relatively vulnerable infrastructure nodes or assets that had never been designed to be resistant to serious attack. Much infrastructure was viewed and developed historically as a public good and was easily accessible to those that might want to disrupt networks. At the same time the US security services, especially the military, were also acutely aware of the vulnerabilities of cities' infrastructure as the United States was actively developing and using new weaponry for controlling and disrupting telecommunications, electricity, and transportation networks (Graham 2006). Consequently the issue of urban infrastructure moved up the security agenda, especially in the United States.

In particular networks of cities and the critical roles in global infrastructures, especially as hubs and nodes on international networks, were acknowledged. That is the key communications and transportation infrastructures that connect major global cities through complex aviation, shipping and telecommunications networks (Graham 2006) as hubs and that are usually co-located closely within large global cities. These infrastructure networks themselves were seen as potentially risky, and in the case of 9/11, the very infrastructure had been used as a weapon against large symbolic urban targets in New York and Washington. Consequently there was critical concern that these networks themselves were vulnerable to terrorist or military attack and were particularly difficult to defend because the defining logic of their governance had been to open and extend them through privatization and liberalization and also to facilitate greater degrees of interconnection and flows between large cities and nation-states. Yet increasing attention, particularly in a US context, focused on how the perceived vulnerabilities of these networks could be addressed through national and international action in the governance of aviation and shipping. The networks that facilitated high levels of flows of people, resources and goods were now seen as being possibly a source of network vulnerability potentially affecting the operation of the global economy (Graham 2006).

Additionally and importantly, there is the tension between, on the one hand, infrastructure networks that are mainly in private ownership and operation, that have been working within a logic of privatization and liberalization to facilitate easier, more efficient and less costly flows of *selected* people goods and services and, on the other hand, the perceived vulnerability of infrastructure networks that have not been explicitly designed to build in resilience or security and whose governance is highly fragmented over different nation-states, public and private sectors and a range of associated actors and interests. Consequently a critical issue for CIP has been how to develop infrastructure protection strategies at the scale of large city-regions and through international networks when there are not necessarily existing governance frameworks. In response to these pressures a shift has taken place in CIP with a refocusing on the role of cities, especially the largest world cities and the national/international networks connecting them. Increasingly CIP effectively categorized and focused on the most 'important' urban locations and on addressing perceived weaknesses in the vulnerability of infrastructure networks, particularly the shipping and container port system linking cities.

The US Homeland Security Department distributed $844 million in fiscal 2008 grants for protection of ports, transit systems and other components of the nation's critical infrastructure. 'More than $333 million went to the 8 *highest-risk* metropolitan areas in the Urban Areas Security Initiative' (this included the Los Angeles/Long Beach Area, San Francisco Bay area, Washington National Capital area, Chicago area, Jersey City/Newark area, New York City area and the Houston area). The 23 *second tier* metropolitan areas received just under $17 million.[1] In each of these high-risk metropolitan areas studies and strategies were developed to understand the risks and vulnerabilities of infrastructure and to develop governance capacity that could coordinate an appropriate response.

'Natural' disasters and environmental security – the emergence of urban resilience

What is of interest for us is how the initially narrow conception of CIP at the urban level – being conceived in terms of vulnerabilities and risks of infrastructures to terrorist and or military attack – has been enlarged to a wider conception of urban resilience. Urban resilience is concerned with the ability of cities to develop the social and technical capacity to deal with shocks and crises. Rather than a narrow notion of physical security it looks at how cities can develop the preparedness to deal with disruptions and threats, of a multifaceted nature, to return the city to 'normal operation'. This has not totally displaced the securitized or terrorist focus but has become layered over it in complex and sometimes contradictory ways.

The turn to urban resilience has been underpinned by a series of emergent pressures which we explore here. A series of natural disasters in the United States and elsewhere – including Hurricane Katrina, the north east blackout of 2003, and so on – rehighlighted that many CIP issues were not only related to terrorist activity but also closely linked to other forms of disaster and vulnerability. These incidents had serious implications for critical infrastructure which was vulnerable to disruption because of the cascading effects of failure. These were particularly stark in New Orleans where police were unable to coordinate an effective response due to the failures of power and communications systems. A shift took place in debates about physical security to wider questions about the capacity and capability required to prepare for both natural and terrorist disaster. This addressed the social and institutional issues in emergency planning and the development of more effective and efficient strategies for coping during an emergency and post-disaster reconstruction. Improvements focused on organizational and institutional innovations to develop resilience at an urban level. Although the terrorist response elements were still critical, preparedness focused on a much wider set of potential threats to the city that encompassed flooding, heatwaves, cold and ice, earthquakes, severe weather disruption, and interruption to infrastructures due to natural and technological breakdown.

The shift in developed economies towards questions of resilience also echoed a much more long-standing debate within developing cities' contexts which had been concerned with understanding systemic vulnerabilities of especially low income populations to 'natural disasters'. Here evidence clearly showed that for structural and economic reasons low income and the poorest communities often occupied the most vulnerable and high risk locations. Debates about improving resilience were echoed in the developed cities, while in developing cities' contexts there was increasing focus on a wider set of vulnerabilities, often associated with a lack of infrastructure, such as health, drought and so on. Layered over these developments were more upstream concerns that began to become integrated in the discourse around urban security. For instance, once emphasis shifted from immediate physical protection to questions of natural disaster this highlighted wider questions about how vulnerability and resilience may be constituted and understood. For example, within a developing city context long-standing questions about water stress and food security at the urban level were being echoed in a developed city context. The security of water, food and even energy supplies started to become part of wider debates about urban vulnerability and resilience.

These more upstream concerns were accelerated by new issues such as the implications of climate change – partly in terms of how climate change may impact on major urban areas – through heat island effects, changed weather risks and so on but also second order effects about how this might lead to shifts in urban populations and impacts on resource security. The interest of military and security services in the potential impacts of climate change also started to reshape urban questions about resilience which then pushed into wider questions of environmental security.

Urban economies: carbon regulation, competitiveness and low carbon innovation

National states have a critical role in securing longer-term economic reproduction and national economic competitiveness and consequently the implications of climate change are becoming reinterpreted within wider state roles in relation to the economy, innovation, technology policy, markets and jobs. In particular climate change is becoming layered into the wider economic role of state policies as an ecological and economic form of territorial state regulation. The national state starts to develop new territorialized targets and inventories for national carbon control and regulation, systemic change in socio-technologies of infrastructure which are then cascaded down on to cities and regions. What starts to happen here is that new forms of carbon control then place pressures on cities and regions to develop new strategies and systemic responses. The ecological becomes selectively incorporated into wider questions about the continued economic reproduction of cities and regions.

Carbon control and regulation have potentially significant implications for urbanism, where

> [e]xplicit carbon quotas will open up a new phase of socio-spatial regulation, and far sighted regional and local decision makers are busy preparing for the new era by developing carbon-reduction strategies and modelling the complex carbon impacts of different spatial planning and management trajectories.

One consequence of this is that '[s]ome authorities, such as the Greater London Authority, are seeking to secure their post-competitive future by experimenting with alternative energy supply, investing in low carbon infrastructure and ensuring that present planning decisions will not be a burden in the future' (While 2008: x–xi). This is happening against a backdrop of international attempts to build a consensus around the need for reducing greenhouse gas emissions and in particular carbon dioxide which are likely to have significant implications for cities and regions. Recognition of the importance of developing responses to climate change mesh with the role of national governments in setting voluntary and mandatory carbon reduction targets that are likely to be 'cascaded' down on to subnational territories – such as regions, subregions, cities and localities/municipalities – below the national level. These targets, which may be statutorily enforced by national and even international agreements, produce a new ecological pressure on cities and regions.

Many western nation-states have developed carbon reduction targets ranging, for example, from Norway's plans to cut carbon emissions to zero by 2050 to the UK's national target for reducing emissions by 80 per cent on 1990 levels, also by 2050. In response to these targets there has been increasing interest in climate change, low carbon lifestyles and the development of carbon budgets. It is evident that carbon considerations will impact on many aspects of contemporary life. There has been much consideration of the implications of carbon control for places focusing on action to reduce carbon footprints and developing low carbon energy transitions, based on shifts away from fossil fuels towards renewable energy technologies, greater energy conservation and efficiency and new transport solutions. All these issues have been, to some extent, on urban and regional policy agendas for a decade or more but are likely to be given even sharper focus by the development of carbon targets, quotas and emissions trading explicitly in relation to places. This new logic is likely to have significant implications for urban and regional regulation.

Carbon regulation: the emerging 'spatial logic' of control

Synonymous with the cascade approaches to environmental governance that we discussed in the previous chapter the logic of carbon control is that once the global emissions-reductions requirement is agreed, 'it is then translated into a series of territorially-based targets organised at the scale of the nation state' (While 2008: ix). This would mean, for example, that the nation-state could devolve and cascade its national targets into binding quotas for regions, subregions and city-regions and local authorities/municipalities at the subnational level.

Cities and regions – with their variable carbon emissions (see Figure 2.4 for example) – may be given statutory carbon reduction targets, like, for instance, national UK targets for local authorities to reduce emissions associated with their premises and estates (Department of Energy and Climate Change (DECC) 2009). A critical consequence of such targets is that they give much greater significance to carbon reduction strategies as continued economic growth is dependent upon a territorial ability to manage carbon emissions downwards in order to meet national targets. In this scenario low

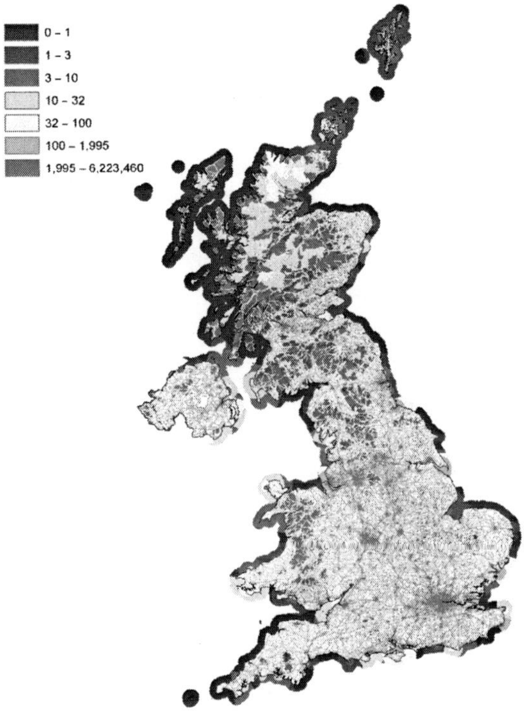

Figure 2.4 UK emissions map of carbon dioxide.
Source: Defra.

carbon development and transitions in the social and technical organization of the built environment and infrastructure may become strategic to place-based growth ambitions.

If targets are developed as part of national and even international carbon control regimes, there may also be the opportunity for the exchange and trading of carbon credits between different places. There are a number of possibilities, for example, of funding projects to reduce carbon emissions in other places or even countries in return for additional carbon emissions associated with growth strategies. Another option is cap and trade schemes where a financial price is attached to carbon emissions which can then be traded as commodities. Those that save carbon emissions are rewarded by trading with those that exceed their quota and who have then to pay for their pollution by buying additional credits. Such schemes have already been developed within Europe for large companies and have even been proposed for individuals. They could also be extended to places and localities. The benefits for government of such targets and quotas are considerable in that they provide certainty about the target or quota but leave it up to others including places to decide how to respond and meet their own targets.

Securing competitive post-carbon futures: emerging urban implications

This new spatial logic is likely to have critical implications for cities because they are major producers of carbon emissions. There are two critical issues for cities in considering the potential implications of carbon targets and quotas. First, there is the political process of how targets and quotas are allocated to cities. Second, there is the issue of how cities manage longer-term social and economic priorities within the context of carbon quotas. We will briefly consider both these issues.

Many sources suggest that cities are responsible for 75–80 per cent of all greenhouse gas emissions (Stern 2006)[2]. Given this assumption there are serious strategic implications of carbon quotas for cities such that ways of reconciling growth ambitions against a context of carbon constraint would need to be found. Yet the role of cities in producing carbon emissions is seriously contested. By contrast to the 75–80 per cent figure, the IPCC's fourth assessment shows that carbon dioxide from fossil fuels accounted for 57 per cent of world anthropogenic greenhouse gas emissions in 2004 and a very large proportion of non-carbon-based emissions (methane etc.) are not located in cities – agriculture and gases released from land use changes contributed 31 per cent of all greenhouse gas emissions (IPPC 2007).

There are particular problems with existing assumptions that cities are producers of 75–80 per cent of greenhouse gas emissions (see Satterthwaite 2008). While there are detailed carbon inventories for major cities in the United States and United Kingdom, there are too few cities for which this is the case. This being the case, the figures of 75–80 per cent reported by the mayors of New York and London are not necessarily transferable to other cities. In addition to this, it is often very difficult to assign greenhouse gas emission activities to a particular city. So, for example, emissions assessments could be allocated to where goods are produced or where they are consumed. Urban authorities obviously have a strategic interest in how emissions are allocated particularly given the differences that result from allocating emissions between where they are 'produced' and where the resources are 'consumed'.

A productionist approach, focused on the location of where the greenhouse gases are emitted, leads to a line of argument that 'using no more than a (hopefully educated) guess' cities generate between 30 and 40 per cent of anthropogenic greenhouse gas emissions. In this scenario emissions that take place outside cities – such as power stations, mining and manufacturing – are not included in cities' emissions despite such goods and services being consumed within cities. Also transport-related emissions are not assigned to cities, even if a high proportion is tied to the movement of goods and people in and out of cities. So for instance, the debate about where to allocate sea and air related emissions means that New York and London's carbon inventories are presented with and without air- and sea-related emissions (Satterthwaite 2008).

A consumption-based approach, focused on the places where the carbon emitting goods, services and resources are consumed, would result in cities being assigned a higher proportion of the emissions than the productionist logic – estimated at being no more than 60–70 per cent of total emissions (Satterthwaite 2008). In this scenario greenhouse gas emissions associated with the production of goods, services and energy consumed within the city is assigned to the city (even if it is produced in, for example, China or India). This approach includes the methane generated by cities' wastes but located outside the city and also emissions associated with goods that are then sold outside the city would be subtracted from a city's emissions.

Each method produces quite different results. Comparative work on Beijing, Shanghai and Tokyo showed that the two Chinese cities had higher per capita emissions when considering the emissions produced within the city but that Tokyo had much higher per capita figures if the inventory includes emissions associated with goods and products purchased within the city but produced elsewhere. Consequently, the largest, most developed world cities that import large amounts of goods and resources are powerfully affected by these different measurement techniques.

Although these two methods produce quite different results, we need to be aware that they mask significant differences between and within cities in their role in greenhouse gas emissions nationally and internationally. Many of the world's largest and wealthiest cities would not be among the world's highest greenhouse gas emitters because most electricity and energy intensive goods are imported (see Dodman 2009). For example cities such as New York and London would not figure among cities with the highest emission per person in their national context because they have little heavy industry, import most of their electricity and in part have lower levels of private car use and less energy intensive housing.

Most cities in low-income nations will have far lower levels of greenhouse gas emissions per person within their boundaries compared with high-income nations irrespective of what technique is used. These can vary per person by more than a factor of 100. If it was possible to adopt a consumption perspective, cities that concentrate wealthy people, with high-consumption lifestyles would have emissions per person that would be thousands of times higher (Satterthwaite 2008). Of course, as we have seen above, most of the cities at risk from the impacts of global warming are in low and middle income nations where the low income populations are most at risk. There are also large and significant variations in per capita figures for greenhouse gas emissions within cities so assigning emissions to cities misses the very large differences in emissions between individuals and households. It is the poorest households, for example, that have the smallest and even negative emissions (see Hardoy et al. 2001).

Given the discussion above, critically the development of subnational carbon targets and budgets will be of critical interest to large cities for many reasons. In particular, how carbon emissions are developed and the techniques used will have critical implications for cities as the different methods produce such varying results. Of key significance will be whether there is a focus on consumption or production related approaches. Whether aviation and sea transport are included or excluded is likely to have profound implications for the largest global cities where these infrastructure are co-located. Related to this is the critical issue of how quotas will be set for different localities, by whom (international, national or regional levels) and with what consequences for particular cities and populations within them. So, for instance, if aviation is excluded from a large city like London's quota does this give it greater comparative advantage?

A further set of issues is how carbon targets and quotas might affect strategic decision making in cities. Difficult choices about what are the most important developments and investment decisions might have to be made in the context of a constrained carbon budget. Cities would have to develop their own 'carbon control fixes' within the overall target assessing how the management of carbon associated with waste, energy and mobility could be more effectively managed. In this context we can begin to understand the interest in some cities in low carbon or carbon neutral development and attempts to develop the required capacity and capability to build zero carbon buildings, districts, blocks and cities that could be a competitive advantage against a background of carbon constraint. The capacity to effectively manage carbon could then become a key metric in city comparisons and economic development strategies.

The implications of these pressures leads to wider concerns about uneven development and spatial justice and raises the issue as to whether these pressures will lead to a new more progressive form of development – and, in doing so, link to traditional urban sustainable development concerns – according to a wider set of criteria that includes ecological concerns. Alternatively, another view might be that this leads to more intensive competitive pressures between place-based interests in different cities as they seek to protect and insulate themselves from carbon pressures. This may result in prioritizing some sectors and people over others and seeking to gain competitive advantage through an ability to develop technical fixes or have their cities prioritized in carbon quota allocations due to their national and international importance. With these possibilities in mind the issue becomes one of which cities and social interests might be better placed to develop the relations and capacities to adjust more effectively to post-carbon futures than others.

Conclusion

In this chapter we have examined how an emerging set of broad issues in relation to climate change and resource constraint have created a broad range of pressures and impacts on existing relations between cities, infrastructures and resource flows. This we outlined as a series of urban threats, an extension of a military and political discourse around infrastructure security and the recasting of a national economic competitiveness discourse and the ways in which these are creating pressures for and informing of the selective reconfiguration of relationships between places, infrastructures and resource flows.

Urban threats from climate change impacts we detailed as being manifold and variable in their consequences. We also pointed out that there are significant gaps in our understanding of these threats and impacts given the limited number of city-specific studies that have been undertaken and the ways in which extrapolations often follow from these limited studies. We examined how critical infrastructure protection has begun to be extended into wider sets of questions about the vulnerability and resilience of infrastructure systems more widely in an urban context. In particular the ways in which such debates are starting to push into discussions about the longer-term consequences of climate change for energy, water and food security issues in an urban context; where through a state security discourse urban infrastructure and climate change are being reinterpreted as a political and military question for the national state and cascaded down on to cities and regions.

We also assessed the ways in which we are seeing a shift from a global debate about carbon control to a more spatialized logic that develops carbon targets and quotas at urban and regional levels. This rescaling of the logic of carbon control to cities is significant given that cities may be responsible for between 40 per cent and 70 per cent of carbon emissions depending upon the calculation methodology.

This last point makes a priority of place-based capacity in cities to respond to carbon targets and quotas which may affect their ability to compete with each other for growth. When set alongside attempts to understand the impacts of climate change and resource constraint at a city-scale and the extension of an infrastructure security discourse to incorporate ecological dimensions, this is likely to create the conditions within which urban responses to these pressures are produced. What types of responses this leads to is the subject of Chapter 3.

PART II

Urban Responses

3

Rebundled urban network ecologies

Introduction

In this chapter we focus on developing an understanding of the scope of urban responses to climate change and resource constraint. We are interested in trying to comprehend how different types of urban responses – new development projects, experiments, initiatives, strategies, policies and infrastructures – are being constructed as attempts to respond to climate change and resource constraint. In particular we review how different urban interests – urban authorities, development companies, engineering consultancies, planners and architects, utilities and infrastructure providers, corporate and energy companies – are developing new types of responses. We draw on a wide range of documentary and representational sources to seek to understand four types of emerging responses.

Initially we review the emergence of new types of *integrated eco-urbanism* that through urban development projects at a range of different scales – houses, villages, towns and cities – seek to integrate demand management and local production technologies within the architectural and engineering design of new-build projects to develop more resilient and self-reliant types of urbanism.

We then examine efforts to develop *systemic urban transitions* in the existing city through the use of new concepts such as low carbon cities, zero carbon cities, peak oil cities and so on. These are simultaneously new societal visions of contemporary urbanism and also set up new technological expectations about how cities, buildings, infrastructure, institutions, publics and stakeholders and so on might be involved in a wider system change in response to new ecological pressures.

Following this we move on to focus on the development of *new urban network technologies* that attempt to address climate change and resource constraint. Here we are particularly interested in the efforts in cities to develop alternative energy networks – hydrogen, biofuels and district heating and cooling schemes – grey water and recycled water schemes, that are retrofitted into urban contexts partly to develop new infrastructure systems and that are partially designed to deal with the implications of climate change and resource constraint.

Finally we review new strategies primarily focused on the *reconfiguration of existing infrastructure* networks in the city. In particular these are attempts to reshape particular networked technologies in a specific way and are primarily focused on water, food, and energy and flood security. Here the conception is on preparing an urban infrastructure network to deal with the implications of changes, ecological and economic conditions, through systemic attempts to reconfigure the resource flows, technologies and aspects of the social organization of network infrastructures.

Critically we address these responses that are being developed and suggest that they seek to 'rebundle' together customized, localized, integrated and enclosed 'network ecologies' of infrastructure networks and critical resources. This is a rebundling into bounded assemblages that attempt to build and seek to guarantee the long-term security of resource provision in relation to particular customized spaces – neighbourhoods, blocks, towns, islands, cities and regions. We conclude the chapter by arguing that there appears to be a critical shift in efforts to rebundle infrastructure and ecology through different scales of urban enclosure that is exemplified by four new 'types' of infrastructural response that attempt to rebundle together ecology, infrastructure and urban developments in very particular reconfigurations.

New urban developments as 'integrated eco-urbanism'

A set of emerging responses are primarily concerned with attempts to construct *integrated* responses to infrastructure that cut across multiple infrastructure networks – energy, food, water, waste and so on – and that are rebundled together at particular scales in the design of new buildings, neighbourhoods, towns, blocks and cities. These usually focus on new-build developments, either entirely new 'greenfield' developments such as an eco-city or eco-town or sometimes new standalone developments that are located adjacent to or within existing cities such as an eco-house or eco-neighbourhood. This style of development is much more concerned with integration at the scale of the development than with the wider transformation of the existing city or its incumbent infrastructure networks.

These responses have at their core the vision and aspiration that they are able to transcend conventional notions of ecological constraint – climate change and resource constraint – as they build ecological security by internally producing their own food, energy and other critical resources, reusing wastes as resources and reducing reliance on external infrastructures. The examples we detail below and synthesize draw upon the most 'exemplary' illustrations of this new style of urbanism that are claimed by their developers to offer *the* new and replicable models of development. We allow the developers – the consultancies, engineers, architects – to describe their concepts and their replicability through their own words, statements and representations of space. A pattern starts to emerge within which particular coalitions of social interests – consultancies, architects and engineers sometimes with elements of the green movement – are collaborating with particular place-based interests in the development of new infrastructural fixes.

Eco-towns

Eco-towns are a programme of supposedly exemplar sustainable new towns to be built in England. The wider context within which eco-towns development sits is in the UK government target to build 240,000 new homes per annum by 2016 and to reduce carbon dioxide emissions by 80 per cent below 1990 levels by 2050. Consequently it is intended that eco-towns 'should make a significant contribution to these targets and help address the serious threat of climate change [and] should be exemplar projects that encourage and enable residents to live within managed environmental limits and in communities that are resilient to climate change (Department for Communities and Local Government (DCLG) 2009: 2).

In 2007, the UK Department of Communities and Local Government announced a competition to build up to ten eco-towns. Initially over fifty bids were submitted and in mid-2009 the government eventually gave the go-ahead for the construction of four eco-towns, providing 10,000 new houses. Eco-towns are meant to reach new ecological standards that are 'more challenging and stretching than would normally be required for new development' and are designed to act as 'exemplars of good practice' (DCLG 2009: 1).

More specifically this relates to the intention that in eco-towns, energy emissions specifically related to the built environment will need to be 'zero or below'; development will need to be 'resilient to and appropriate for the climate change now accepted as inevitable'; they are ambitious in terms of water efficiency and in areas of water stress and should aspire to water neutrality – that is build new developments without increasing overall water use in the wider area by managing demand downwards; and that they should consider using locally generated waste as fuel sources for combined heat and power regeneration for the eco-town (DCLG 2009: 3).

Eco-Blocks

Eco-Blocks have been developed as an urban gated community that is 'resource self-sufficient (i.e. carbon neutral) in its operation (or close to it), and if it could replicate and spread throughout the world, this would be a major force in reversing global climate change' (Fraker 2006: 44). The Eco-Block project has been developed as an alternative to China's typical 'gated super block' development model which is and has been dependent on central infrastructure support in the form of electric power plant generation and distribution, sewage collection and treatment, water treatment and distribution, and waste collection and disposal.

China is currently hard pressed to satisfy the infrastructure demands of the super block model. In contrast, the Eco-Block is designed in an integrated whole-systems approach to generate all its energy from on-site renewables, to recycle all of its water and to recycle over 80 per cent of its waste for on-site uses. In addition, it is designed to provide convenient pedestrian and bike access to a new bus rapid transit system located on a major adjoining road network. The designers claim that the Eco-Block's whole-systems approach is flexible and adaptable to multiple local conditions and climates and is 'widely replicable throughout China'.[1]

The aim is that the 'neighbourhood becomes essentially a self-sufficient unit, a circular system. It does not require the construction of expensive new power plants, new sewage treatment and new water supply outside the system' (Fraker 2006: 49). Although the block does not have to be connected to the grid for energy and is self-sufficient in water the waste cycle is not totally closed and approximately 20 per cent of waste will need to be disposed outside the system. Arup, the engineering firm, are now partnering the University of Berkeley to develop an Eco-Block as a demonstrator with the stated intention of replication of this throughout China.

Eco-islands

Eco-islands provide a different scale, again, at which attempts are underway to construct more self-reliant forms of urban development using decentralized technologies and recirculating resource flows. Islands offer a number of benefits in developing these types of solutions; in particular, that there is a clear boundary that provides a high degree of clarity about the direction and scale of resource flows; often governance structures may be more unified and simplified so that infrastructural priorities can be more easily designed into the scheme, implemented and monitored; and there is a defined range of publics and stakeholders to engage with on new infrastructure solutions. Consequently islands have traditionally been used as sites of experimentation for a number of infrastructure experiments, such as the testing out of the conversion to natural gas and piloting water metering on the Isle of Wight off the south coast of England.

Treasure Island is a former naval air base built on 400 acres of reclaimed land in San Francisco Bay in the 1930s (Figure 3.1). By 2020 it is planned to become one of the most sustainable communities in the United States with 6000 new apartments and businesses. Over 50 per cent of the power, it is claimed, would be from renewable resources, including solar electricity and solar water heaters, the street grid has been designed to maximize the exposure of the rooftop photovoltaics to sunlight and all the buildings would be within a 15-minute walk of a ferry terminal to San Francisco. It is intended that residents could obtain much of their fresh produce from a local

Figure 3.1 Treasure Island.
Source: ARUP.

organic farm that would use as fertilizer the waste from a water treatment plant already on the island.

Even so, the development would not feasibly be net carbon zero given the tensions between the available technology and the need for the owner to make a profit. There are, however, other aspirations to design zero carbon islands. The island country of Tuvalu, for example, located halfway between Australia and Hawaii in the middle of the Pacific Ocean and – with most of the country only one metre above sea level – highly vulnerable to climate change and sea live rise plans to abandon fossil fuels and generate all of its energy from renewable sources by 2020.[2]

Eco-cities

In 2005 Arup, the global planning, engineering and design consultancy, signed a contract with the Shanghai Industrial Investment Corporation (SIIC) to plan the world's first sustainable city – an eco-city – at Dongtan, in Shanghai, China (Arup 2005). Dongtan is situated in a strategic position very close to Shanghai and on the third largest island in China, located at the mouth of the Yangtze River. Dongtan is three-quarters the size of Manhattan and is designed as a sustainable city to attract a whole range of commercial and leisure investments to what is currently a large area of mostly agricultural land (Figure 3.2). The aim is that all Dongtan's buildings are powered by renewable energy and that it is self-sufficient in water and food sourced from the surrounding farmland. The 630 hectare first phase of the city aspires to house up to 80,000 people by 2020 and inhabitants will be encouraged to make use of the zero carbon public transport, which will be powered entirely by renewable energy (Arup 2006). The Arup director, Peter Head, has claimed that the scheme 'could well prove to be the *template for sustainability in city planning* [author emphasis] – not only in China, but elsewhere in the world' (Arup 2005). A wider eco-cities movement, including Masdar, in the United Arab Emirates (UAE), and Chula Vista in

Figure 3.2 Dongtan 'the world's first sustainable city'.
Source: worldarchitecturenews.com.

the United States, has also developed with the intention of developing more autonomous and self-reliant types of urbanism.

Eco-regions

The Thames Gateway is a £9 billion project that would span 40 miles along the River Thames from London's Canary Wharf to Southend in Essex and Sittingbourne in Kent (Figure 3.3). The UK government has also repeatedly said that it wants the project to be an example of responsible environmental development to the rest of the world and consequently the Thames Gateway is proposed as an 'eco-region' using new environmental technologies and higher environmental standards than previous UK development projects.

The eco-region designation includes energy savings modifications with the aspiration of reducing the carbon emissions of existing homes in the region by 60 per cent. Newly built homes would be completely carbon neutral. It is also intended that the project should be water-neutral through the application of an intensive water management system that is to be put into place in both existing and new homes (Environment Agency 2007). This would include the use of water-saving appliances, variable-flush toilets,

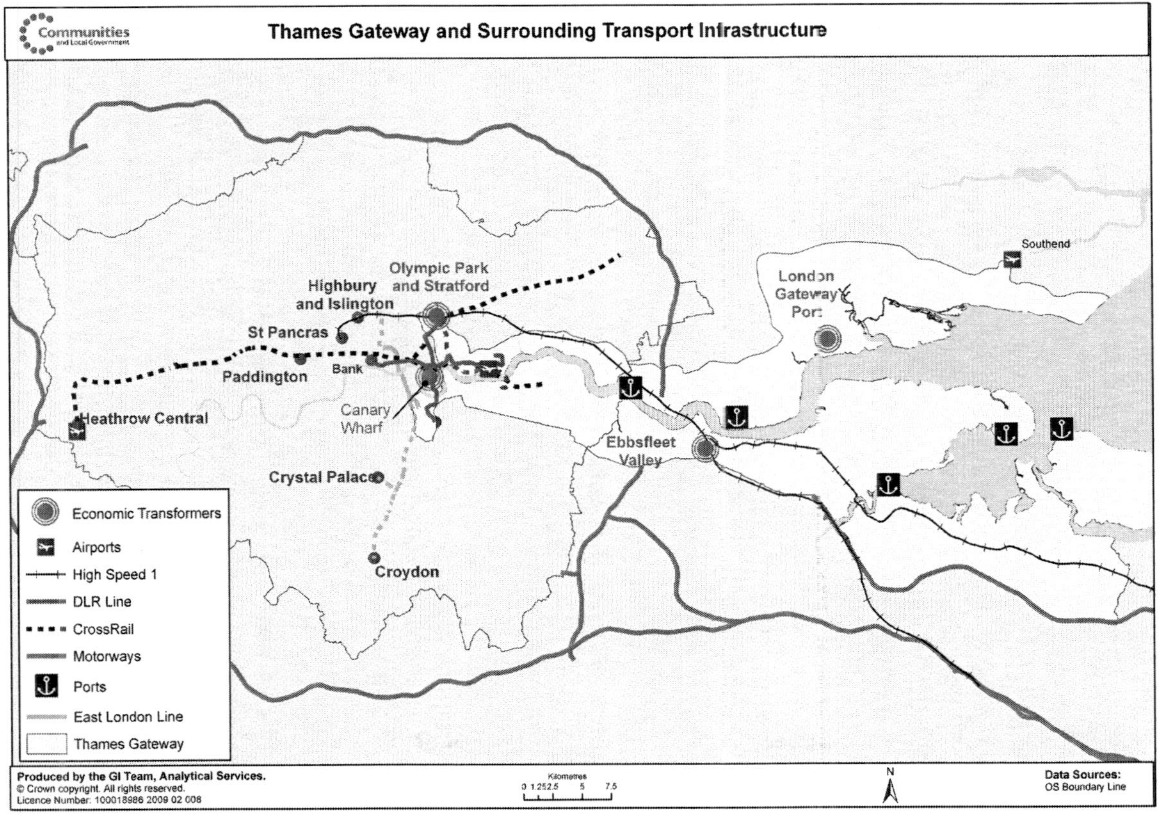

Figure 3.3 Thames Gateway 'eco-region' near London.
Source: DCLG.

low-flow showerheads, and low-flow faucets. Additionally the proposal includes plans for what would be two of the largest offshore wind generation dams in the world and a biomass generation plant for energy. Thames Gateway would also be incorporated into the Olympic site plans for 2012 and Cross Rail, a proposal to expand the current UK rail system with a new high-speed underground railroad across London and South-East to be completed by 2017.

Low Impact Urban Development

Low Impact Urban Development (LID) encompasses a range of community-based initiatives that seek to internalize infrastructure and resource flows. LID is important as a site of practical innovation and attempts at low carbon living (Pickerill and Maxey 2009). Although there are important similarities between LIDs and the more commercially and governmentally oriented integrated eco-developments outlined above – in particular the emphasis on autonomy, the development of local technologies, circular metabolisms and the aspiration for greater self-reliance – there are also some significant differences. In particular LIDs stress local and community control of infrastructure and raise wider issues about ensuring more equitable access to environmental resources for low income households.

Constructing new 'urban networked technologies'

In this section we examine responses that are primarily concerned with the development of alternatives to conventional energy and water networks through the construction of new infrastructure systems and the creation of new or restructured resource interdependencies. The aim of these developments is to build more resilient resource flows, at the scale of the city, under conditions of climate change and resource constraint. In the energy sector the emphasis on alternatives is underpinned by strategic interest in the development of new energy networks at the urban scale and beyond including district heating systems for the distribution of heat and cooling, and alternative fuel networks such as hydrogen, biofuels and other gases. In respect of water it is focused on the development of parallel water systems for the distribution of grey water and recycled water to sit alongside potable water systems. With transport there is enhanced concern with the development of parallel mobility systems in the city such as electric car mobility systems, cycling networks and the application of biofuels and hydrogen in vehicles. Underlying all these responses is the attempt to construct new or parallel infrastructures within the city.

New energy networks

New biofuel, hydrogen and district heating and cooling energy networks and systems are important features of urban responses to the challenges posed by resource scarcity and vulnerability to climate change, but they are also predicated on new potential sources of economic growth. A critical feature of these responses is that they are based on renewable energy sources such as biofuels – ethanol and methanol but also waste and biomass – and can be utilized as a fuel for electricity generators, transport or heating. They can be organized as combined heat and power (CHP) systems or combined cooling, heating and power (CCHP) systems. The effect of decentralizing energy systems is that transmission efficiency is increased and heat generated in the production of electricity is also utilized rather than wasted.

US cities have been particularly active in the use of biofuels, made from plant matter or from recycled fats, oils and grease that have been collected from restaurants and fast food outlets, which is then blended with or directly substituted for gasoline or diesel. Although a range of urban social interests view biofuels as a way to reduce dependence on foreign oil, improve environmental health at the local level by reducing air and water pollution and stimulate jobs growth innovation and economic development, there is virtually no discussion among US mayors, however, of the negative impact of US biofuel production on world food prices. Consequently 'the well-meaning goals of US mayors for energy independence and economic development may contribute to the hardship and even hunger of millions around the globe'.[3]

In San Francisco 100 per cent of the city's vehicle fleet is now powered by biofuels and Las Vegas, Honolulu, and Fort Wayne, Indiana, for example, are among several cities moving rapidly towards the exclusive use of biofuels in municipal vehicles and equipment. New York City Mayor Michael Bloomberg's environmental initiative requires that one-third of the heating oil purchased by the City contains biofuel. The growth in urban biofuels is driven by US Federal research policies and tax credits that incentivize the production of biofuel manufacturing facilities. Biofuel manufacturing plants are frequently located on the outskirts of cities, equidistant from the agricultural areas that supply the raw materials and the urban consumers of the final products. For example, six new or proposed plants will surround Seattle, and three will ring Rochester, New York. Within a European context cities are collaborating through the Biofuel Cities European Partnership in order to demonstrate the use of biofuel technologies and sustainable mobility policies.[4]

Hydrogen cities are city-level strategic responses to the challenges for cities posed by peak oil and climate change. Hydrogen cities often, either explicitly or implicitly, acknowledge the extent to which urban economic and everyday life is predicated on oil, coal and gas and see hydrogen, produced from renewable sources of energy such as solar and wind, inform the development of community grids and city-level mobility systems. This move towards becoming more 'self-sufficient' is part of a strategy of prioritising the preparation of cities to address the challenges of peak oil and climate change. It is a strategy of ensuring that cities, with their dependence on fossil fuels, can protect their economic positions. Attempts to translate

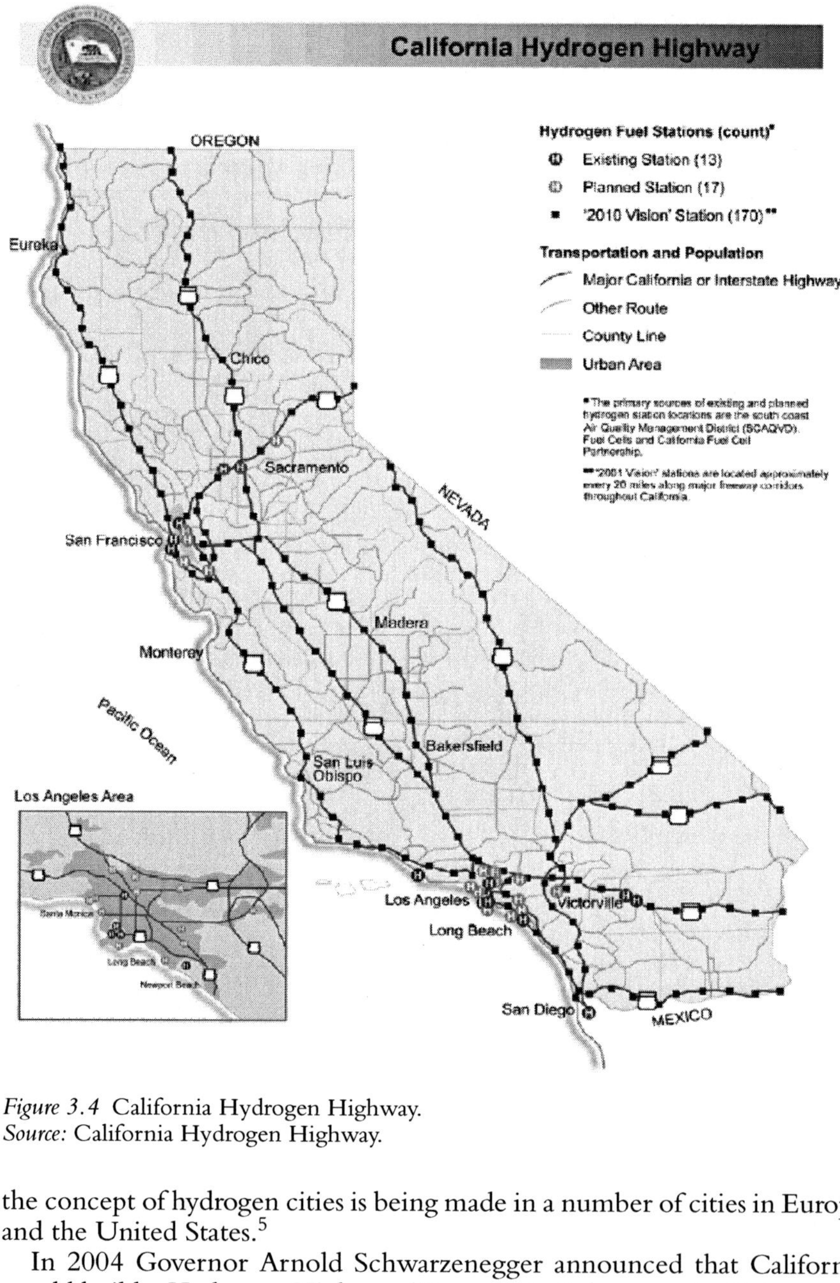

Figure 3.4 California Hydrogen Highway.
Source: California Hydrogen Highway.

the concept of hydrogen cities is being made in a number of cities in Europe and the United States.[5]

In 2004 Governor Arnold Schwarzenegger announced that California would build a Hydrogen Highway (Figure 3.4). Subsequently, 21 interstate freeways were designated as the underpinnings of the California Hydrogen Highway Network. The objective is that the California Hydrogen Highway Network should achieve a 30 per cent reduction in greenhouse gas emissions compared to fuels and vehicles in 2005 (California Environmental Protection Agency 2005: 3). The goal of the transition to a hydrogen transportation economy is to reduce dependence on foreign oil and 'to prove to the world that a thriving environment and economy can coexist.[6] The first phase of development is based on plans for 50 to 100 refuelling stations, servicing 2000 hydrogen vehicles by 2010, with the stations initially clustered in the San Francisco Bay area-Sacramento region and the Los Angeles-San Diego region.

New water networks

Innovations in urban water management are increasingly seeing water 'holistically' where all water is 'recycled' water.[7] In this view sources of water are interlinked with first use sources, including run-off in catchments, groundwater and harvested rainwater are interlinked with second use sources that include the use of industrial and domestic wastewater, agricultural drainage and sea water. The consequence of viewing water holistically has been not only to encourage water efficiency but also to promote strategies for using wastewater and innovative ways of sourcing, treating, storing and discharging water in order to more effectively deal with drought and longer-term uncertainties about the security of water resources with a changing climate. Second use sources include not only wastewater but also urban stormwater and desalinated sea water. Rainwater recycling utilizes rainwater from roofs to use in washing machines, toilets, for watering gardens, cleaning cars and so on. Grey water recycling uses shower and bath water for use in washing laundry, flushing toilets and so on. In an urban context the Sydney Water Reclamation and Management Scheme (WRAMS) demonstrates the systemic urban use of new water networks that utilize innovative technologies. As Australia's first large-scale urban water recycling scheme its aim is to save more than 850 million litres of drinking water annually through providing adjacent sporting venues, businesses and a community with recycled water. As a system for water management numerous processes and technologies constitute WRAMS. These include a water reclamation plant that removes water from sewage, water storage reservoirs, a water treatment plant to filter and disinfect water from the water reclamation plant and storage and a separate, dedicated supply system to pipe water from the treatment plant through Sydney Olympic Park and Newington.[8]

New transport networks

Alternatives to conventional transport networks are being developed in a number of different ways, including through the promotion of electric cars, urban cycling schemes and hydrogen fuel cell bus networks. In the case of electric cars, in 2009 plans were announced for nine UK cities and towns to be involved in an £11 million development plan to have charging points for electric and hybrid fuelled vehicles.[9] At a different scale, the

Mayor of London, in claiming to make London the 'electric car capital of Europe', aims to have 25,000 charging points and 100,000 vehicles across London by 2015.[10] The Clean Urban Transport Europe (CUTE) initiative, to take another example, was established at the end of 2001 and was primarily a coalition of supranational political interests (the European Commission) and multinational automobile (Daimler-Chrysler) and energy (BP) organizations. CUTE involved the demonstration of 27 fuel cell powered buses in nine European cities (Amsterdam, Barcelona, Hamburg, London, Luxembourg, Madrid, Porto, Stockholm and Stuttgart), alongside associated projects in Reykjavik and Perth (Figure 3.5). These emerging transport networks are not solely confined to electric cars and hydrogen buses. In July 2007 the Velib public cycle hire programme was launched in Paris. The system comprises over 20,000 cycles and more than 1400 hire stations.[11]

Community networks

Alongside more corporate-dominated or policy-led responses there are a set of community responses to the development of new water, energy and transport networks. Taking the case of energy this includes initiatives such as community energy cooperatives, community renewable energy collectives and so on. The Community Power Fund (CPF), for example, is a community-owned financing organization the aim of which is to support community-owned renewable energy in Ontario.[12] Similarly, the Hearthmakers Energy Cooperative is a cooperatively owned and community-based business that works to provide renewable energy and environmental services and products to members in the Southeastern Ontario region.[13] Next Generation Energy Alternatives (NGEA) is a non-profit, membership-based corporation which aims to rent roof space on condominiums, housing co-operatives, and other large buildings where photovoltaic (PV) panels can be mounted to generate clean power. Electricity from these panels will be sold to the electricity grid at 42 cents per kilowatt hour for 20 years under Ontario's Standard Offer Contract programme.[14]

Alternatives to corporate and city-led initiatives are frequently based on efforts to broaden the range of participation in developing responses. Community energy networks, as well as being about new networks of hardware, also deal with the provision of a forum for mutual self-help. In that respect they are concerned with producing networking opportunities for a wide range of individuals, societies and community groups interested in reducing energy consumption and making savings and how they can help each other.[15] These networks are often concerned with learning and the sharing of resources, knowledge and contacts.

Reconfiguring cities as 'systemic urban transitions'

In contrast to the development of new networks there is also an emphasis on the development of systemic responses to the reconfiguring of existing cities

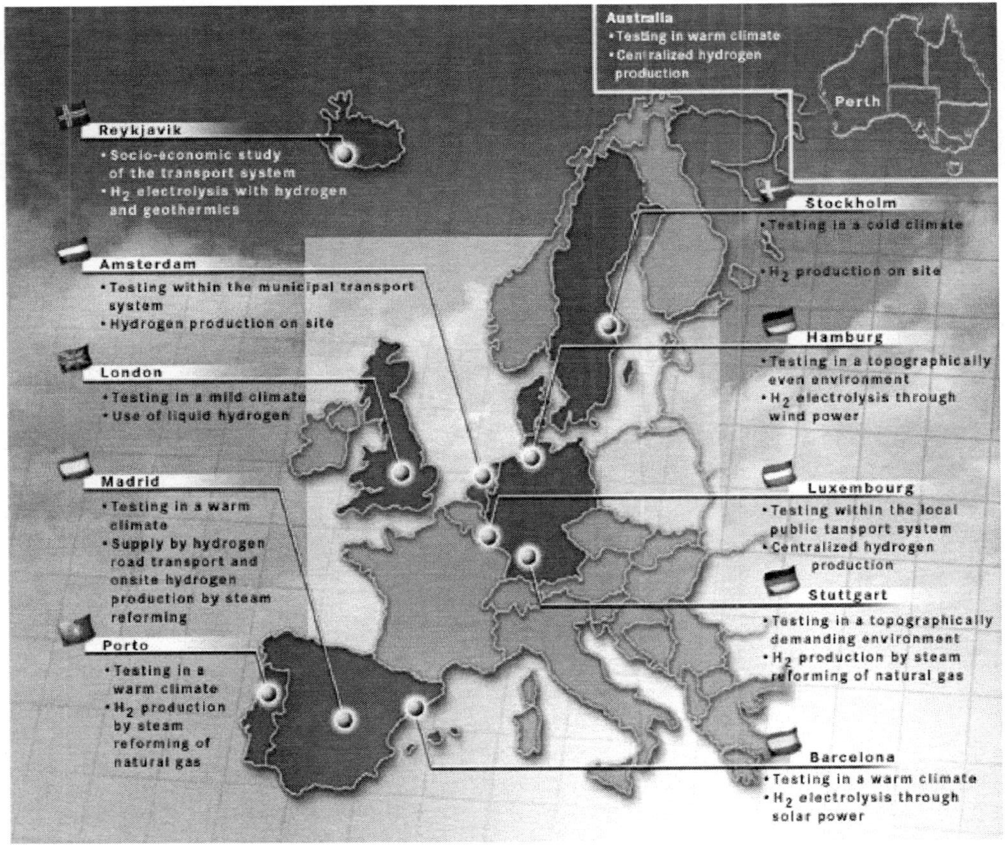

Figure 3.5 New hydrogen transport networks.
Source: fuel-cell-bus-club.com

in relation to *packages of infrastructure*. Such developments represent attempts to develop a concept of purposive urban transition in the socio-technical organization of cities and existing infrastructure systems. Critically this means a concern with the social, institutional, political and technological complexity of reshaping cities' existing networked ecologies. These are often developed by groups of cities' leaders and decision-makers, researchers, developmental and international agencies. These sorts of strategies imply systemic socio-technical change in the organization of cities in order to prepare for a period of climate change and resource constraint. As such they are usually underpinned through the development of wider social visions about the type of city that is being constructed, wider forms of engagement with 'stakeholders' and publics about the construction of the vision – although the depth and scope of this may be variable. The cultivation of a strategic orientation for the reconfiguration of socio-technical systems – infrastructure, buildings, social relations – also requires the purposive, strategic development of new forms of knowledge, capacity and capability to translate these into action.

Low and zero carbon cities

Low carbon cities and zero carbon cities envision the reconfiguration of cities in ways that are significantly different from their existing organization and use. This is an acknowledgement that cities currently consume large amounts of carbon-based resources and, as such, new lower or zero carbon networks need to be developed. Reconfigured cities aim to utilize and make better use of existing resources through reusing wastes, using renewable and decentralized sources of energy and attempting to make cities more 'self-sustaining'. This in turn means gaining more local control over resource flows as an important component of future economic growth. In doing this there is an important role for new green technologies to improve the efficiency and effectiveness of resource flows and also in promoting behavioural changes through redesigning cities in terms of low and zero carbon housing, organizing cities less around the private use of cars and more in terms of public transportation, walking and cycling. In low and zero carbon cities there is an emphasis on high quality public spaces and on the design of streets in a greener way.

Low carbon cities are being developed in different national contexts across the globe. Trends of rapid urbanization, heavy industrialization and related resource consumption and carbon emissions mean that China offers a useful example through which to view the opportunities of low carbon cities. The Low Carbon Cities Initiative (LCCI) aims to 'explore low carbon development models' (see Figure 3.6),[16] by supporting the research and the implementation of policies which aim to contribute to low carbon development, exploring new finance and investment instruments and sustainable trade opportunities for energy efficiency and the renewable energy industry, and improving public awareness on climate change in order to enable and encourage Chinese citizens to save energy and the environment.

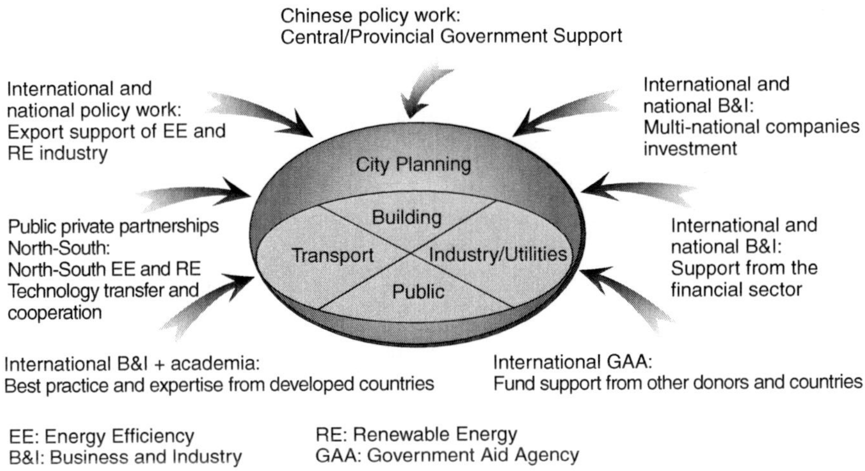

Figure 3.6 Conceptual framework and key drivers for the low carbon city.
Source: WWF.

Initial pilot projects will be in Shanghai and Baoding. In Shanghai this will include: piloting new eco-building demonstrations; improving the energy efficiency of existing buildings; and engaging with publics in promoting energy efficiencies and savings. In Baoding activities will include: developing networks of sustainable energy in relation to knowledge and technologies; encouraging investment in and export of sustainable energy products; and design and implementation of a Solar Energy Demonstration City action plan and a Wind Energy Industrial Park development plan. In doing this various social interests are coming together in coalitions of interests. These include: banks (HSBC), trade associations (Chinese Renewable Energy Industries Association; Chinese Wind Energy Association), research interests (Shanghai Research Institute of Building Science), municipal government agencies (Shanghai Construction and Communication Commission) and the World Wildlife Fund.[17]

Peak oil cities

Peak oil cities or post-oil cities attempt to develop systemic alternatives to cities' dependencies on oil. Now that production of oil may have peaked, or is likely to peak shortly, there is both uncertainty and potential vulnerabilities across different urban contexts. There is urgency in preparing cities for a post-peak oil era and energy constraint as a means of addressing both the short-term aspects of peak oil (e.g. shortages) and longer-term economic vulnerability (see Lerch 2007).

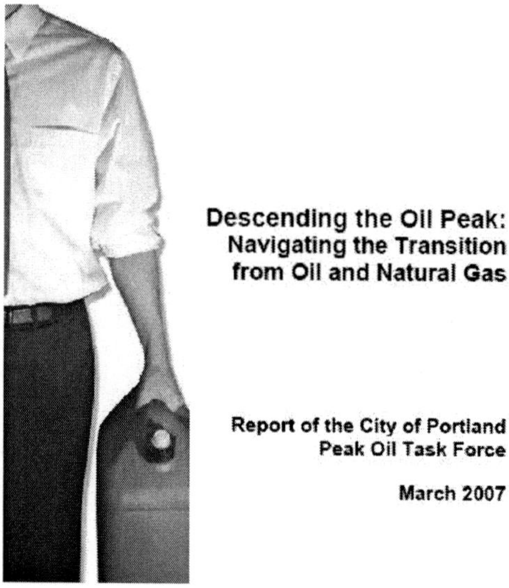

Figure 3.7 Peak Oil City Portland.
Source: portlandonline.com

To address this vulnerability peak oil cities are committed to significant reductions in oil and natural gas consumption, for example Portland, Oregon, following a 2007 resolution, plans a 50 per cent reduction over the next 25 years (Figure 3.7).[18] This would require the transformation of energy and food systems of production and consumption. In doing this there is also recognition of the social stresses of peak oil and its economic implications by encouraging the strengthening of community cohesion social and economic support measures. In medium-sized cities in the United States, such as Portland, Bloomington, Dallas and Austin there is acknowledgement of the need to address the significant social and economic vulnerabilities, such as unemployment, that result from peak oil. The perceptions of vulnerability among these cities, in the US context, has been underpinned by work on this agenda among local governments in the face of frequent federal refusal – under the presidency of George W. Bush – to engage with a post-oil and greenhouse gas reduction agenda.

Additionally Hurricane Katrina – and the subsequent fuel shortages it resulted in – forced many local governments to be aware of their vulnerabilities given the limits of federal and state government assistance. In this context there was recognition of a particular role for local government in building post-peak oil resilience through its powers, particularly in relation to transportation and planning, through mixed-use developments and also given its proximity to these issues (Lerch 2007). There was also widespread recognition that use of what are essentially limited local government responsibilities to build resilience and relocalization requires adept use of limited resources and also the effective building of relations with stakeholders.

Transition towns and relocalization movement

Community-level responses to the mitigation of both peak oil and climate change are also being undertaken with the intention of developing urgent and planned action at a local level to build the resilience that is deemed to be missing in a globalized, oil-reliant economy. The Transition network was founded in 2005 and consists of small towns mainly in the UK but also including larger cities such as Bristol and Nottingham and communities in New Zealand, the United States, Australia, Canada and Germany. Transition Towns are built on the basis of moving away from a dependency on global systems of energy, food, transportation and health production and consumption and – through the relocalization of these systems – building up local resilience and independence (Hopkins 2008). The Transition network questions the pervasive vision of unrestricted global growth and, in response, promotes the mobilization of local capabilities to develop particular local visions of what a Transition Town looks like. This takes place within the parameters of a wider generic Transition framework and through the development of plans for building community resilience (Figure 3.8).

Doing this generally involves a small initiating group who utilize the Transition Model to seek to engage local people and develop a Transition Initiative. This approach involves a significant process of building community-level awareness of peak oil, climate change and the challenges they pose and the opportunities they present. There is a need to develop a

Figure 3.8 Transition Towns.
Source: deeprgreen.co.nz

multiplicity of relationships and connections with: existing groups across the community, particularly in relation to areas such as food, energy, transport, health, economics and so on; with local government; with other transition initiatives to develop understanding and share lessons and support; to develop and implement projects aimed at developing community-wide understanding of resilience, carbon issues and community engagement. The aim is that this eventually informs the development of a long-term (15–20 years) community implemented Energy Descent Action Plan.

Retrofitting existing urban networked infrastructures

In this section we primarily address urban responses through a specific infrastructure network that attempts to systemically reshape key critical aspects of the social and technical organization of the system in response to climate change, resource constraint and other pressures. These responses are being constructed and developed by urban interests – policy, research, activist – through waste, energy and water networks at the urban scale, and represent attempts to systemically reshape existing infrastructure networks in order to reduce vulnerability, increase self-reliance and develop resilience. These tend to be infrastructure-specific strategies designed to address issues such as water security, energy security, food security, to develop more flood resilient cities, to make use of urban waste flows in order to extract resources and energy and to exploit the potential of smart technologies and pricing systems to reconfigure the use of existing infrastructures.

Urban water security

Security of water supplies is increasingly a significant issue at an urban scale (for a review, see Cisneros and Rose 2005). In the context of serious drought and water restrictions, Australia offers the emblematic example of urban response to water scarcity and ensuring water security. More specifically, the Australian government, through its *Water for the Future* plan, provides broader parameters with which state and territorial responses are formulated. A significant part of this response has been not only the promotion of technological responses, demand-side management and efficiency programmes, but also and in particular the championing of desalination plants.

The National Urban Water and Desalination Plan, as a component of *Water for the Future*, has promoted the creation of centres of excellence in desalination technology and water recycling, major new urban water recycling projects and proposals for the construction of a desalination plant in Adelaide.[19] This mixture of recycling, building desalination plants and development of the knowledge and relevant technologies is an attempt to 'drought-proof' Australia's drought-hit cities. While recycling measures may be relatively uncontroversial, desalination plants have attracted huge levels of controversy. This has been on cost grounds but primarily has related to the huge amounts of energy required to power desalination, the discharge of concentrated brine into the sea and the potential ecological consequences of this for sea life. This tension and struggle between economy and ecology can be seen in a major new initiative – the Urban Water Security Research Alliance – that is seeking to

> develop new employment opportunities and the freeing up of economic growth from the increasing uncertainty of rainfall and climate change, closing the water cycling loop and reducing dependence on the environment to supply water and assimilate waste and contributing to the adoption of new practices in urban wastewater and storm water recycling to increase our national water security.[20]

Urban closed material loops

In the waste sector there is interest in the potential of developing closed-loop production systems that can minimize the cost of waste disposal and of moving raw materials between different sites. Clearly these types of developments work best in large densely populated cities and regions with proximate, complementary partners to balance resource flows, with close institutional connections to reduce transaction costs and regulation to encourage collaboration rather than the disposal of by-products as wastes. Although raw materials still have to be brought into the city and finished products exported outside the region these types of initiatives represent attempts to relocalize and look for opportunities to remobilize waste by-products as commodified resource flows through closing the loop in production processes.

The Kalundborg eco-industrial partnership in Denmark is emblematic of attempts to construct eco-industrial development. Five industrial companies collaborate for mutual economic and environmental benefit through closing material loops by exchanging different kinds of waste by-product (see Figure 3.9). Wastes that might have been deposited in the environment thus form valuable inputs into the production process of spatially adjacent companies. These new interdependencies and resource exchanges attempt to close material loops by recycling, reuse and recovery of wastes through common systems of integration.

Flood resilient cities

Since the 1980s rapid urban development and climate change provide a wider context within which there has been a general increase in the frequency and severity of river floods (Serre and Barroca 2009).[21] This has challenged the levels of protection afforded by existing flood defences, where failure of these defences can lead to increased flood risks of particular urban areas in times of extreme weather. Existing assessments, forecasts and

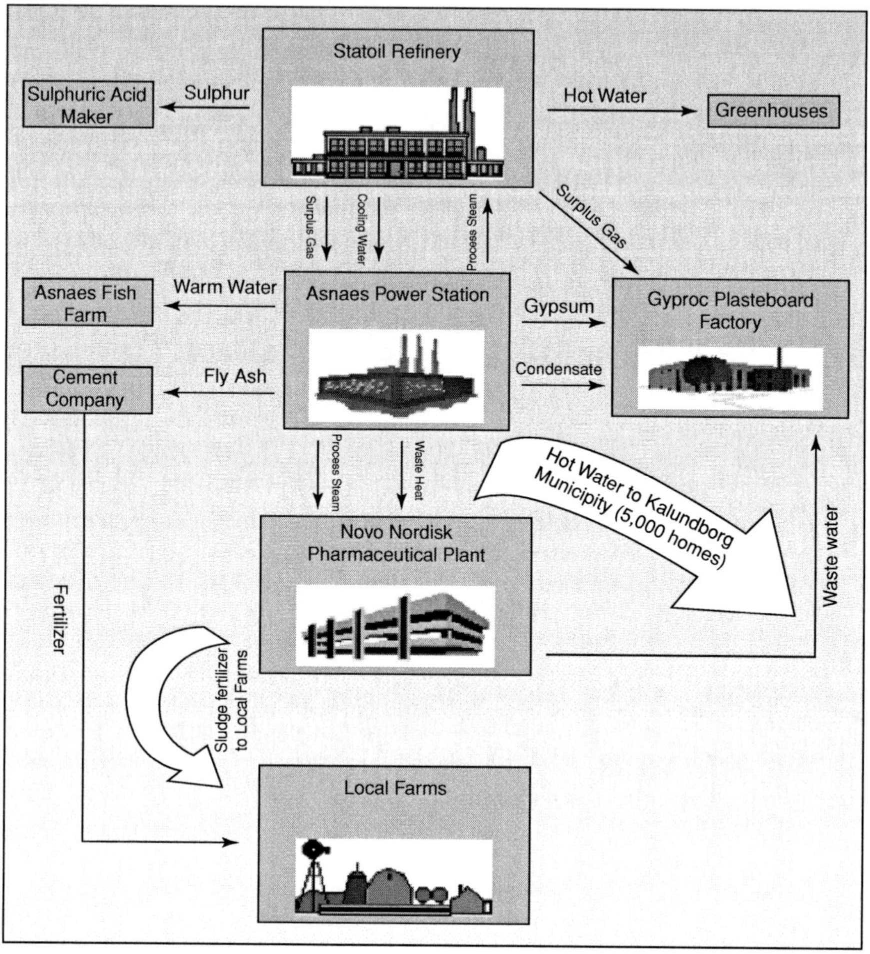

Figure 3.9 Symbiotic relationships between the core partners of Kalundborg.
Source: ipts.

modelling of flood risks have often been seen as reflecting inaccuracies and producing partial accounts of urban flood risk. Furthermore, traditional flood-risk management approaches often omit important ways of dealing with floods proactively at city level.[22] The development of the idea of flood resilient cities is recognition that in an era of increasing urban economic and population growth and pressures from climate change-related impacts flood risks to cities are uncertain. Given this uncertainty existing flood defence arrangements are being questioned. Building flood resilient cities through urban responses to flood risks encompass a range of adaptation and mitigation measures. This includes what are deemed to be the appropriate policies, regulations, institutions and levels of funding to underpin the development of flood resilient cities.

A knowledge base and an understanding of the potential impacts of vulnerability to urban flooding at a local scale are required in respect of flood probability and its risk to urban infrastructures and the built environment. This necessitates modelling and scenarios of climate and flood impacts and consequences.[23] Such assessment and analysis is used to underpin the decision-making process of which forms of response (the types of technologies, the allocation of financial resources, public perceptions and so on) are thought to be required in particular urban contexts to enhance cities' resilience to flooding. The issue this raises is whether and how attempts to build flood resilience are built into strategic plans, how the strategy is communicated and to whom in building effective capacity to act.

Urban energy security

Attempts to build urban energy security are increasingly informing strategies of energy independence at the level of the city (Figure 3.10). In San Francisco, for example, a coalition of Local Power, the Sierra Club, Greenpeace USA, Our City and a grouping of San Francisco community groups produced the Community Choice (CCA) Implementation Plan. The backdrop to the Plan is the passing of legislation by the state of California that permitted its cities and counties not only to switch electricity power suppliers for householders and businesses, but also to fund renewables and energy conservation projects as a means of reducing dependency on existing natural gas and nuclear plants. The California Public Utilities Commission made the decision to allow cities such as San Francisco to break away from their existing relationships and contracts with utility power providers and to take responsibility for their own energy future.[24]

At the level of San Francisco, the Energy Independence ordinance directed City departments to prepare an Implementation Plan to solicit 360 Megawatts of new solar photovoltaic installations, distributed generation such as fuel cells, wind turbines, hydrogen, energy efficiency, and conservation technologies as standard components of the City's electricity service. On an average day, San Francisco requires 650 Megawatts of power at night and 850 Megawatts during the day. A 360 Megawatt investment in green power – as called for in the City's adopted Electricity Resource Plan – will far exceed the Renewable Portfolio Standard called for by state law (Local Power 2007).

Urban food security

Urban food security has been a long-standing research and policy issue in cities of the global south (for a review, see Koc et al. 1999). Now debates about urban food security have become part of the food-related policy interest in cities and communities of the global north. Greater uncertainty

Figure 3.10 Energy independence for San Francisco.
Source: Local.Org

about the safety and reliability of food production systems has generated significant interest in urban and regional closures and economic resilience around urban food systems. The urbanization and regionalization of food systems has been promoted by a variety of factors including healthier diets, organic production, shortening supply chains and local production. The enclosure of food may reduce the ecological foot print of agriculture through reducing food miles and carbon dioxide emissions as supply chains shorten. However, this may not necessarily be the case. If we take the example of producing tomatoes in heated greenhouses this may generate more carbon dioxide emissions and water consumption than producing them at a distance from their final consumption. Agriculture systems might be made more ecologically efficient by locating food production factories near to city-regions, through lower transport costs, rebalancing supply and demand and responding to local needs but this would require careful configuration and integration of energy, water, waste and logistics systems.

Increasingly though local food production is being linked to questions of ecological security. A number of urban food movements have started to make explicit links between local food production and national security.[25] For example in the United States it has been argued that greater connections need to be made between homeland security priorities and the development of a safe regional food supply that is less vulnerable to external threats. In order to prepare for emergencies, the suggestion is that all cities and communities should be able to 'produce or supply at least a third of the food required by its residents. At present, less than 5 percent is produced'.[26]

There are many issues about whether local food production does actually increase food security for a city, with the risks that, potentially, reliance on a single source may actually increase vulnerability. Instead it may make sense to rely on a more widely distributed and more varied supply system. However, despite these questions new-designs for eco-cities are increasingly emphasizing local food production attempting to find ways of squeezing such systems into the urban fabric.

A typology of rebundled network ecologies

Figure 3.11 provides a typological characterization of these responses that is organized around two dimensions.

The first of these dimensions – aligned on the horizontal axis – is whether urban responses focus on 'new build and new construction' of buildings and new networked infrastructure or whether urban responses are concerned with the 'retrofitting' of existing cities and already installed networked infrastructures. The importance of this distinction between new and existing is two fold. First, in the case of new build developments – such as homes, commercial, office, industrial development, industrial parks, new neighbourhoods etc. – it is relatively more straight forward within the control of the site boundaries to develop innovative infrastructural solutions. Although new infrastructure networks need to be squeezed into congested urban cores the social and physical challenges of developing new infrastructure are more complex than site specific new build but generally less complex and costly

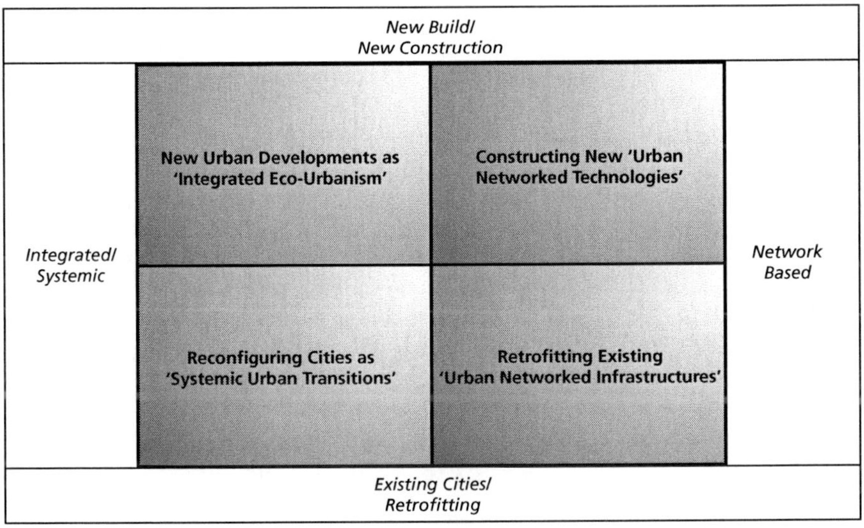

Figure 3.11 Four types of rebundled network ecologies.

than retrofitting and existing infrastructure. This leads us on to the second distinction. A critical challenge for cities in the global north where renewal of the existing built fabric and infrastructure is relatively slow – usually a few percentages per annum – is how to manage change in existing cities built fabric and infrastructure. Consequently attention here shifts to the complexity of existing infrastructure and built environments that have to be retrofitted to reduce resource use and implement decentralised technologies. This will involve multiple users, landowners and building ownership as well as incumbent and competitive utilities, commercial, public and community interests, regulators, governments and many other competing social interests. Thus, the first dimension is therefore the distinction between new build and retrofitting – with too much attention focused on the former and less on the latter.

The second dimension – represented in the horizontal axis – is whether urban responses are concerned with the development of integrated or systemic change at the level of new build and existing city or alternatively are they mainly networked focused – that is, concerned with particular infrastructures rather than the wider understanding of the city. The significance of this distinction is three fold. Let's first take the example of network based solutions. These primarily focus upon either shifts in the socio-technical configuration of existing infrastructures or the development of entirely new infrastructures located within the city. These tend to have a much more infrastructural focus more concerned with narrower technical changes in the configuration and design of networked technologies primarily steered and led by infrastructural providers and developers. The second response concerns integrated/systemic developments at the scale of new build developments, ranging from buildings and neighbourhoods to islands, cities and regions. Here integrative and systemic responses are designed into the eco-development prioritising autonomous and self-reliant resource flows. Finally the third response concerns reconfiguring existing cities built fabric, infrastructures, publics wider stakeholders and social interests around wider visions of systemic socio-technical change – post-peak oil cities, low carbon cities etc. This implies a much more ambitious, challenging and complex process of managed socio-technical and behavioural change in how a city relates to its key infrastructures and resource flows. These responses are likely to be more complex involving much more blurred boundaries between different resource flows and social interests that cross multiple scales. Systemic change within a highly distributed infrastructure system is much more demanding and difficult to achieve whether we are talking about the installation of new networks in congested cities or the reconfiguration of existing systems.

Within each of these types there is what we characterize as 'alternative' responses developed predominantly by environmental and/or community groups. These usually are outside the more corporate and policy oriented solutions, generally have a less technologically focused emphasis, or where technologies are involved they tend to be more focused on demand-side management and localised and small scale production technologies. There is a greater emphasis in these responses on social, cultural and behavioural change than technological solutions and a more household or local focus.

Conclusion

In this chapter we detailed and illustrated a range of urban responses to climate change and resource constraint. We situated these wide-ranging developments in relation to the spatially variable rebundling of infrastructure, ecology and security. In doing this we developed four types of response in relation to rebundled networked ecologies. These are visions that develop a set of expectations about new types of urbanism that significantly reshape cities' relations with resources through changes in the social organization and scale of socio-technical systems that support the contemporary city. Running through these visions and aspirations are a number of issues. Not least of these is that cities can systemically anticipate and prepare for a period of resource constraint and climate change. This logic seeks to identify and respond to different types of resource constraint – for example water stress, flooding, energy security, carbon constraint – in order to ensure that cities can continue to extend, expand and reproduce themselves.

Furthermore, within these new infrastructure fixes – self-sufficiency, decentralized technologies, local production, circular metabolisms and so on – is the view that resources and infrastructures can increasingly be bounded at the scale of the city. This implies a withdrawal or reduced reliance on existing regional and national infrastructures as cities become the locus for attempts to build up the capacity of infrastructures and supply resource locally.

Finally the social interests involved in these initiatives – engineering, consultancy, design and architecture, banking and governmental interests – all explicitly claim that such models are exemplary and replicable in other contexts and may become the defining logic of urban development. A select number of places – San Francisco, London and Dongtan – at least claim, and are positioned as, those cities producing the new infrastructure fixes which are then consumed by other cities. Such a shift represents a significant change in how cities relate to resource and infrastructures. Rather than a logic of continued extension as cities seek out critical resources the new logic at least implies a greater degree of self-sufficiency as cities attempt to bound and localize resources. Given these issues, now we examine how these types are being mobilized in relation to the world's largest and most economically powerful cities.

4

Secure urbanism and resilient infrastructure in world cities

Introduction

This chapter examines the ways in which urban infrastructural innovations developing in world cities are represented and enacted. World cities have an emerging and increasingly influential role in developing responses to the challenges of resource constraint and climate change through coalitions of interests actively positioning them as the context for action.[1] Political leaders in world cities are working together, primarily with multinational corporations, to develop network responses to new pressures for securing the material reproduction of their cities and also building resilient infrastructures.

Critically, relationships between the world's largest cities and resource flows are being reshaped in ways that are leading to new configurations of infrastructure provision. In order to better understand these processes of reshaping, we examine a wide range of documents and draw on conversations and discussions from our interviews with urban and national elites and decision-makers, with representatives of utilities, regulators, multinational corporations and environmental justice organizations. In constituting responses, ensuring the strategic protection of cities from the impacts and effects of climate change and resource constraints becomes a central element. Underpinning this are investments in understanding the city-specific and long-term effects of climate change, especially in relation to flood risk and temperature rise, and the development of strategic flood protection systems, green infrastructure, and retrofitting to deal with increased temperatures.

World cities are simultaneously contexts where attempts are being made to build greater degrees of 'autonomy' and 'self-reliance' into the supply of water and energy, the mobility of people and goods and the disposal of wastes. Traditionally the prosperity of cities has been secured by seeking out resources and waste sinks from ever more distant locations. Yet this approach is now being reversed as cities seek to become more self-sufficient by reducing reliance on international, national and regional infrastructures and reinternalizing their own resources and the recirculation of wastes. Urban elites and multinational corporations are also collectively building agglomerations of new urban mobility systems within and between world cities. At the same time as focusing on local 'enclosed' resources, there are also attempts to seek to guarantee intra-city and inter-world-city mobility through the development and use of new mobility technologies, such as pricing, transport informatics and new fuel systems based on hydrogen, biofuels or complex hybrids. We look at each set of responses in further detail below.

Ensuring the strategic 'protection' of cities

Many cities are particularly vulnerable to extreme weather 'impacts' associated with climate change. These impacts include: vulnerability to flooding through sea level rises and storm surges, particularly for coastal cities; an increase in heatwaves predominantly in urban cores with health and increased mortality risks particularly for vulnerable social groups, droughts and reduced snowpack with associated health risks and reduced water availability; and increased extreme weather events such as storms, heavy rainfall and hurricanes, with implications for the existing built environment and for future developments. A critical consequence of climate change is that extreme weather is likely to become more common. Acknowledging this requires that we also recognize that cities are not configured to adapt particularly well to emerging weather patterns and a fundamentally changing climate. One needs only to look at the consequences of heatwaves and floods in recent years for cities and their populations occurring across many cities but exemplified by the Paris heatwave in 2003. The dilemma of a changing climate, with its increased risks of extreme weather, meeting vulnerable urban contexts is increasingly being recognized and strategically responded to by social interests working in, and with representatives of, world cities.

Critically, then, the key response is in ensuring the strategic protection of cities from the impacts and effects of climate change and associated resource constraints. These impacts and effects are likely to be manifold and felt widely within and between cities. Cities are highly but differentially vulnerable to extreme weather events and their consequences and to increasingly constrained resources. Fundamentally many world cities are seeking to build resilience in their socio-technical infrastructures – water, energy, flood protection etc. – and through their capacity to adapt to the risks and impacts of climate change and scarcity of strategic resources (see Figure 4.1).

Central to such strategies are investments in generating a systemic understanding of the city-specific and long-term effects of climate change, especially in relation to flood risk and temperature rise and the development of systemic responses through strategic flood protection, green infrastructure and retrofitting to deal with increased temperatures. Inherent in these

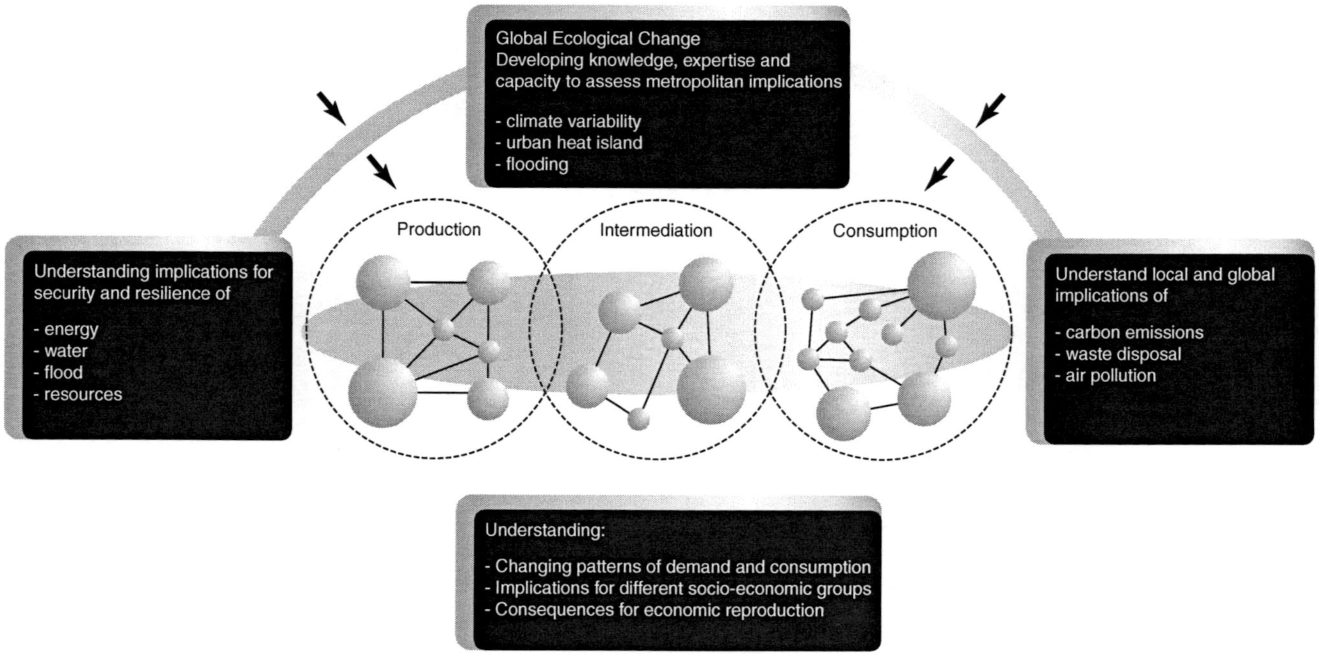

Figure 4.1 Strategic ecological protection of world cities.

responses is the prioritization and privileging of not only particular cities but also particular areas and social interests within world cities – through strategies of preparation and protection that differentially defend, relocate and effectively abandon people, properties and assets.

Strategies of strategic protection focus on the role of world cities as national exemplars and the implications this has for socio-technical reconfigurations. In particular national exemplars are key sites for experimentation, demonstration and development of new infrastructural technologies that are claimed as having relevance and application in other urban contexts domestically and internationally. National government and corporate interests often have a key role in developing these emblematic configurations both within and across different national contexts.

Coalitions of interests in world cities are seeking to ensure that the long-term protection – from heatwaves and flooding – of the city is prioritised within national investment and policy priorities to ensure its long-term reproduction. There are systemic attempts in world cities, including London, Tokyo and New York, to develop strategies, social relations and technologies that are attempting to understand the longer-term implications of climate change and resource constraint on urban reproduction. The Mayor of London, for example, produced a 'world first' strategy to prepare London for climate change (Mayor of London 2008). Critically representatives of world cities are working with national governments and corporate interests to ensure their infrastructure is resilient.

Climate change will impact on flood protection, water, energy, waste, transportation and public health systems that have been configured to cope with less extreme climate-related impacts. Strategic protection approaches recognize the need to identify what the likely climate change impacts are in relation to particular cities, which social groups are more and less at risk, how these are likely to alter over time, and the ways in which socio-technical and behavioural change interventions should be prioritized based on a reading of these risks. Strategies of protection and preparation being devised in relation to world cities are an acknowledgement of the impacts of climate-change-related extreme weather events and also recognition of the need to protect economic assets and properties and to exploit the economic opportunities that lie within strategic responses to these pressures.

Ensuring the strategic protection of world cities is thus predicated on seeking to build resilience and new forms of capacity within socio-technical systems fundamental to cities' economic, ecological and social reproduction. This is a long-term and ongoing process that can be exemplified through particular sets of responses, which include flood risk protection strategies; protecting water supplies and anticipating drought; developing preparatory heatwave plans and cooling cities; and climate-proofing buildings. London, for example, is susceptible to tidal flooding, fluvial flooding from the Thames, flooding from a drainage system ill-equipped to deal with surface water flooding, from sewers and from rising groundwater and from these occurring simultaneously. In addition London's population, economic and development growth aspirations magnify the risk of flooding. The Thames Estuary 2100 study aims to determine the appropriate level of flood protection needed for London and the Thames Estuary until the early years of the twenty-second century.[2] Through the Thames Estuary 2100 project, a series of policy options and 'pathways' are detailed that range from doing nothing to addressing increased flood risk through a series of policies to reduce contemporary and future flood risk. Although London's existing tidal defences offer a high level of protection from present-day flood risks, they were designed to provide 'protection' only up until 2030. After 2030, billions of pounds may need to be spent on raising 300km of other defences to protect the capital – and even more than that if the sea level rises still further (Thames Estuary 2100 2007: 2).

In response, central government in the UK is being actively positioned by the Greater London Authority (GLA) to take prime responsibility for the potential investment required to protect London post-2030 from climate-change-induced flooding. The effects of climate change, such as sea level rise, increased rainfall and storm frequency, mean that London and the Thames Estuary will be at greater risk from flooding in future years. This risk encompasses significant numbers of people and critical assets in the floodplain (for example, 1.25 million people, 69 London Underground stations).[3] Furthermore, many flood risk areas are undergoing development and regeneration, meaning that more people, buildings and infrastructure are likely to be exposed to the risk of flooding in the future. The London Assembly has argued that: 'At least part of this funding would be required to be provided by the Government because of the national importance of protecting homes in and business in the London area' (London Assembly 2005: 10). There is clear strategic prioritization of London within systemic adaptation plans as central government has been positioned as responsible for post-2030 plans and at the same time there is active pressure from the mayor and GLA for the national state to take on the financial commitment of new investment.

The flood protection measures that are being developed in London are mirrored in cities such as Tokyo, Toronto, New York and Los Angeles. These strategies collectively encompass approaches that include the expansion of storm sewer capacity, development of storm surge barriers, revetments, and flood gates, the construction of stormwater reservoirs and rainwater infiltration facilities and the implementation of sustainable drainage systems that utilize permeable pavements, green roofs, stormwater retention ponds and constructed wetlands as well as floodproofing vulnerable buildings.

As well as strategic flood protection the protection of water supplies in a context of increasing extreme weather conditions – particularly drought and storms – raises a critical set of issues for the socio-technical organization of urban water systems. The protection of water systems has multiple facets but is premised on preparation for extreme weather events by improving water efficiencies, reducing losses through system leaks, through the use of new and recycled sources of water and through developing new sources of water. The use of water and water supplies – production and consumption – needs to be adapted to the range of weather events at different times. Wetter winter weather conditions are likely to mean greater run-off into rivers rather than absorption into groundwater supplies. Yet a greater seasonal variation in rainfall will mean that soil moisture will vary with implications for often ageing water distribution networks. Hotter, drier summers, for example, are likely to mean a greater demand for water and in this respect significantly changing and extreme weather events mean that the relationships between water production and consumption and the ecological and socio-technical systems that underpin the production–consumption relationship are in need of adaptation.

Decision-makers in world cities are increasingly seeking to understand the vulnerability, condition and capacity of water distribution and treatment systems. Melbourne, through its *Melbourne 2030* (Department of Sustainability and Environment 2002) strategy is engaged in preparatory plans that seek to protect water catchments and supply facilities and ground water and land resources. Work has also been undertaken to identify and map flood risks, to assess long-term future pressures on existing drainage systems, and to promote flood education and awareness at a community level (White and Howe 2005). In addition, cities such as Seattle, Tokyo and Toronto are working on the promotion of water leakage prevention.

Plans are being developed in Los Angeles, London and New York as a preparatory and protective response to seek to ensure rapid and effective responses to the mortality and health consequences of heatwaves.[4] Additionally they are also promoting measures to keep their cities cooler through reducing urban waste heat and saving energy. The potential consequences of urban heat islands are being addressed through preparatory strategies that include initiatives such as urban greening – street trees, green roofs and so on – creating wind and water corridors and promoting the introduction of renewable energy technologies and also through behavioural change strategies as a contribution to cooling temperatures.

Strategic urban protection also means devising anticipatory strategic responses to extreme weather events and the role of buildings – both existing and new. In respect of new buildings, strengthened codes, designing buildings with more permeable surfaces, with green roofs, improved natural ventilation and raised to address potential flooding is an anticipatory or preparatory form of protection. In already existing buildings retrofitting is promoted.

Building 'closed' cities and decoupling from infrastructure

The reproduction of cities has usually involved seeking out resources and sinks, supported by huge socio-technical assemblages, from locations often ever more distant. Yet this traditional approach is now being challenged as coalitions of interests aspire to reinternalize and relocalize resource endowments by creating closed loops and circular metabolisms (Figure 4.2).

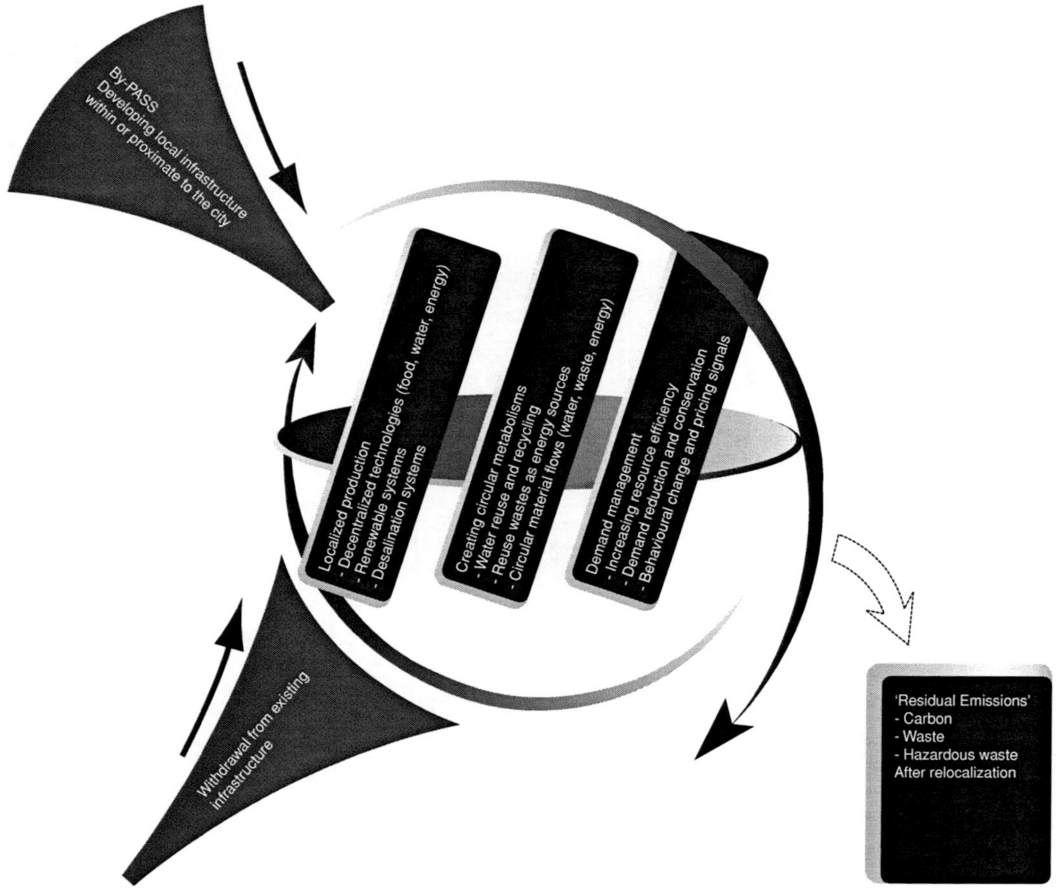

Figure 4.2 Strategic relocalization of infrastructure.

This primarily focuses on attempts to build greater self-sufficiency through socio-technical networks by a dual strategy of both actively disengaging from external reliance on international, national and regional infrastructures and building up local and decentralized systems for water and energy supply, waste disposal and mobility systems. This also envisages an extension of demand-management measures and new pricing mechanisms such as congestion or emission pricing zones. This new approach is no longer based on stretching networks to search out resources more distant from cities but by a strategy of withdrawal and seeking greater independence through developing local resources. Such a socio-technical strategy combines both ecological and security priorities as part of new attempts to guarantee urban reproduction.

The reorganization of resource flows at the scale of the city is part of a strategic attempt to secure the resources necessary to underpin intensified forms of urban growth in a way which positions world cities as not only centres of growth but also a particular form or style of emblematic 'sustainable' growth. Growth becomes strategically intertwined with the (constrained) resources underpinning it and the securing of resources in a context of constraint and carbon emissions reduction. This brings infrastructural innovations and assumptions as to their replicability together with issues of governance, control and the aspirations to bound and enclose resources in an increasingly self-sufficient metabolism to underpin new forms of urban growth. This requires not only the reorganization of resource flows but also transitions in the socio-technical infrastructure and the built environment to support this.

In the case of London, the mayor and the GLA have a clearly articulated strategy for preparing London's infrastructure for major economic, population and housing growth in a style of development that enables London to become the national and international exemplar of a sustainable city. In order to accommodate growth of 800,000 jobs and 400,000 homes by 2016, a complete suite of strategies are designed to (attempt to) guarantee the infrastructural underpinnings for this growth while reducing resource use and carbon emissions. Central to this approach is seeing climate change as an opportunity for developing a new logic of (global) city-regional infrastructure and growth that is an exemplary model for others to follow.

Key parts of this strategy have included developing a strategic understanding of London's resource flows and climate change implications (Tyndall Centre 2008). A suite of infrastructure strategies for energy, waste and water are explicitly designed to minimize the consumption of resources and production of wastes; to consider reuse, develop decentralized energy production and waste treatment technologies; and reduce reliance on external infrastructure to increase the relative self-sufficiency of London. Consequently by 2025 it is the GLA's aspiration that over 85 per cent of waste is treated in London reducing reliance on landfill from 75 per cent to 11 per cent; that energy production is increased to meet 60 per cent of electricity and over 40 per cent of heat demands within London; that inefficiency in water networks has been substantially reduced, through prioritizing the efficiency of networks, reducing leakage, conservation methods and prioritizing water reuse in new development prior to considering new supply options (Mayor of London 2007a). Such innovations are being incorporated into the 2012 Olympics and even the new Thames Gateway project stretching from central London into the Thames Estuary. Here there are a series of initiatives to advance a new form of resource neutral development with commitments to produce water neutral, carbon neutral and waste neutral development (Mayor of London 2006a, 2006b).

This reconfiguration of resource flows in relation to existing world cities is frequently and, as we can see in London, conceptually organized through notions such as carbon neutrality, zero carbon or similar ideas and also through articulations such as energy independence, water neutrality and zero waste. These reconfigurations are premised on a combination of resource conservation and behavioural change strategies and also on replacement sources of energy and water and the technologies to support them. Alongside such strategies, there are attempts to reduce cities' reliance not only through water and energy conservation and waste minimization schemes, but also by developing pricing mechanisms for car-based mobility and reducing reliance on 'external resources'. In addition the world cities where these strategies are developed are not only the context for transformation but also the showcase for these strategies, their operationalization and the range of new technological products.

This we can see as part of a recasting of the neoliberal competitive race to an eco-competitive race. Tokyo Metropolitan Government, for example, in 2007 produced a climate change strategy for the Tokyo metropolitan area (Tokyo Metropolitan Government 2007); a ten-year plan, it went beyond aspirations for carbon neutrality and had as its underpinning the aim of a 'carbon-minus' Tokyo. This is a wide-ranging plan which encompasses energy, water, waste and transportation. In particular, in line with a strategy of building 'autonomy', there are aspirations to reduce energy consumption through demand management measures and also to utilize more renewable energy – particularly solar – as a 'high priority' and waste urban heat. Tokyo aims that energy companies should supply 20 per cent renewable energy by 2020 and, in moving towards this, there is an obligation on energy providers to report on and to produce plans to reduce their carbon emissions and also to increase their share of renewables. This is part of a contribution to a target of reducing Tokyo's greenhouse gas emissions by 25 per cent from the 2000 level by 2020 and establishing Tokyo as a leader in low carbon and low energy technologies.

The combination of reusing energy, energy efficiencies, behavioural change and new sources of energy production are being bundled together in relation to world cities under the auspices of building energy independence and security. This is being done as a preparatory response to resource availability and price volatility, to the challenges of the peaking of oil and fossil-based fuels and also to the vulnerabilities of energy systems to weaknesses, outages and blackouts. Sites of significant energy consumption – for example domestic use, businesses and buildings – become the focus of energy efficiency programmes and also for new sources of decentralized energy. In the case of new buildings the acceleration of decentralized sources of energy is promoted through the mandatory powers held by mayors and urban elites, such as New York's green building law and the London Plan.

The emphasis on energy efficiencies also underpins the relative disconnection from national and regional infrastructures and the development of decentralized city energy systems. In London, for example, the aim is to produce a quarter of London's energy through decentralized systems by 2025 with the majority by 2050 (Mayor of London 2007b). This is anticipated to be a combination of combined cooling heat and power energy and through developing and delivering energy from waste, through small and large-scale renewable energy and will be addressed primarily through planning processes and, in particular a revised London Plan – London's spatial development strategy – to put climate change at the centre of planning considerations and particularly in relation to new buildings with the potential to encourage zero and low carbon developments.

A politics around infrastructure investment is emerging in London. To take one example, when Ken Livingstone was the Mayor of London, he argued that the expansion of decentralized energy generation within London 'will reduce the need for investment in the national grid' (Mayor of London 2006c: 16). However, building decentralized generation within London 'is likely to require modifications to the distribution system within London to ensure energy balancing within London and to enable surplus electricity to be exported onto the national grid' (Mayor of London 2006c: 30). In this vein, large cities are likely to become more active in shaping national investments to try to ensure the potential resilience of the urban grid. The mayor was also active in support of the London Array, a large offshore wind project – outside the boundaries of the GLA – because the power would potentially be supplied to the London grid. More recently, the current Mayor of London, Boris Johnson, has announced his support for a new desalination plant in London to reinforce security of supply in water-stressed London. More widely the water regulators have expressed concern at what they see as London's attempts to secure water supplies without considering wider needs within the southeast of England. At the same time planning policies in London have been used to accelerate the development of water recycling, waste reuse and local energy production technologies in large scale developments over which the mayor can exert some influence.

Actively working towards building greater self-reliance in the urban energy system informs the energy strategy of the City of San Francisco. This involves aggregating existing demand – taken from the incumbent utility, Pacific Gas and Electricity (PG&E) – and renegotiating energy production contracts based on tranches of energy savings within San Francisco, local development of renewables within the city boundaries and the development of renewable resources outside the city. The city has recently been given permission to construct a large direct current (DC) interconnector that would allow it access to renewable resources on its large watershed landholdings way outside the city boundaries and thereby bypass the need to use PG&E's large-scale transmission networks. Such strategies are powerfully resisted by PG&E, but the city authorities continue to work towards the development of strategies to relocalize and where possible explore the bypass of the incumbent networks. Similarly, most of the energy currently consumed in Tokyo is produced outside of the city. The aim is to significantly reconfigure Tokyo's energy metabolism by 2020, with one-fifth of the city's energy consumption aimed to be from renewables by this time and also through additional significant energy efficiency measures.

The same logic of reducing consumption, attempting to reuse resources and develop new sources of production is apparent in relation to world city responses to water constraint. These are exemplified by Melbourne, which is threatened by not only water scarcity but also the increase in such a threat as a consequence of climate change. As part of its response, the City of Melbourne is committed to reducing water consumption by 40 per cent by 2020 and Melbourne Water, the (state-owned utility), to achieve 20 per cent water recycling by 2010. This it proposes to partly achieve through water conservation, including through efficient fittings, appliances and reduced water pressures. In addition, significant consumption of water – through toileting, for gardens, for commercial and industrial purposes – does not require drinking quality water. Through various schemes, including rainwater and stormwater harvesting, residential and commercial processes increasingly utilize water reuse. As part of this process the City of Melbourne recommends such projects are 'climate neutral' in that they do not involve additional production of greenhouse gas emissions (City of Melbourne, n.d.).

In addition to water conservation and behavioural change, the development of desalination plants is being undertaken as part of a strategy of making Melbourne 'drought proof'. An Aus$3.1 billion desalination plant will provide 150 billion litres annually by 2011. It is estimated that the plant will provide around one-third of Melbourne's annual water supply (Victorian Government 2007). There are, of course, questions that arise from these attempts to build a degree of self-sufficiency in respect of water and the potential new interdependencies this opens up in relation to the significant power consumption of the plant. A different emphasis is articulated in Tokyo's plans in relation to water where it is viewed as a 'world leader' in addressing water leakage. Efficiencies in Tokyo's water system are predicated on control of water leakages through a focus on ongoing detection and fixing of leakages. Protection of the water supply system is organized on the basis of maintaining efficiency through prevention and replacement programmes and additionally through water savings technologies and appliances.

World city attempts to reinternalize resources also encompasses flows of waste through processes of recycling, reduction and the use of waste in energy production. Since 2007 Seattle City Council, for example, has had a 'zero waste' strategy that aims to increase recycling, reduce waste and to avoid building new waste management plants.[5] The shift that underpins this is one that moves from seeing increasing waste as obvious and inevitable to viewing it as a resource with a cap on the amount of waste that is exported from Seattle to landfills. The City of Sydney has aspirations for Sydney to become a 'zero waste' community, which again is predicated on an approach that merges the reduction of wastes, with reuse and recycling. In addition to reducing wastes, the claim is that such an approach also saves significant amounts of carbon dioxide emissions and generates revenue. So, for example, a public–private partnership between an Australian company Global Renewables and two Sydney Municipal governments is claimed to save 210,000 tons of carbon dioxide emissions annually and generate US$11.6 million every year. A significant part of this strategy is in the use of a biological digestion and composting process to turn methane to energy without incineration, and in doing so creating 30,000 tons of certified organic fertilizer for farmlands across New South Wales.[6]

Similarly the GLA in London has aspirations to see waste not necessarily as a problem but in a different way as a resource where waste is made 'to work' in London (Mayor of London 2007c) through businesses recycling 70 per cent of their waste by 2020 and the recycling or reuse of 95 per cent of construction and demolition waste. San Francisco, despite its long engagement with waste issues, has an above US average per capita rate of waste generation, this in the context of the United States as a major disproportionate producer of waste,[7] where in the absence of its own landfill capability waste is exported to Alameda County. A significant majority of San Francisco's waste is produced by business and recycling is relatively high. This has led to the development of the view that waste needs to be seen differently where there is a need, in addition to recycling programmes, to avoid producing waste, where this cannot be achieved there should be a move to using reusable packaging and the purchase of products manufactured from recyclable materials. The seemingly paradoxical economic benefits of resource constraints and reduction are mobilized in the context of waste reduction and recycling being seen as major opportunities for local economic growth and revenue generation.

Developing new global urban agglomerations

Seemingly paradoxical to the search for autonomy, new networks of world cities are constituting their interests through initiatives like the C40[8] and the Clinton Climate Initiative (CCI). Coalitions of representatives of world cities and business are increasingly developing responses to the challenges posed by climate change through high-profile infrastructural demonstration projects. What is of significance are the ways in which urban infrastructures are alluded to in these networks but also the positions of particular social, political and economic interests and aspirations that these networks encompass. In the name of world cities new global urban agglomerations of novel mobility systems are being collectively built. At the same time as focusing on the protected space, bounding and enclosing resources, they seek to guarantee intra-city and inter-world city mobility through new technologies including pricing, transport informatics and new fuel systems based on hydrogen, biofuels or complex hybrids.

World cities are working collaboratively to develop new markets for testing and rolling-out global agglomerations designed for world cities in order to guarantee mobility. Emerging socio-technical conceptions of the closed city seem to resonate unhelpfully with the increasing interconnectedness of cities, especially world cities through global agglomerations of communications, airline and international shipping networks. Such networks are responsible for the most rapid growth in carbon emissions and because of the difficulty of dealing with such networks at the scale of a single city such emissions are usually excluded from local policy action.

Yet despite these difficulties many cities are actively engaged in efforts to rethink and selectively experiment with the reconstitution of such global agglomerations. A range of strategies are being adopted including networks between world cities and key corporates to develop new hydrogen-based mobility systems based on clean buses and cars. There are also a range of initiatives around the use of biofuels and hybrid fuel systems to maintain mobility within and between cities. At the same time cities appear to be increasingly strategically interested in how airlines' systems can maintain interconnection for premium business users when there are pressures to address the carbon emissions of the airline sector. Cities are ever more active with corporate partners and national governments in reconfiguring global agglomerations to ensure continued economic and material reproduction (Figure 4.3).

The policy background to these responses is that in October 2005, 18 of the world's largest and 'most important' cities gathered in London under the auspices of the World Cities Leadership Climate Change Summit. The summit, hosted by the then Mayor of London, Ken Livingstone, had the ostensible aim to 'kick-start long-term international collaboration on the pressing issue of climate change' to address 'the inevitable effects of climate change and to exercise leadership in reducing emissions in the future'.[9] At the same time the C20 published its *Low Carbon Leader: Cities Oct. 2005* report (C20 2005), which looked at 15 cities,[10] and where it claimed that the total carbon footprint for these cities was 483 million tonnes CO_{2e} (carbon dioxide equivalent) per year, or about 1.8 per cent of the global total from a combined population of around 65 million. The report also highlighted New York as the city with the highest carbon footprint, Seattle as the city with the highest carbon emissions reduction and Melbourne as the city with the most ambitious greenhouse gas emissions target – zero net from city operations by 2020.[11]

The point of these examples, in the logic of the Large Cities Climate Leadership Group (LCCLG – initially also known as the C20), was to highlight that the interventions and actions of cities show the potential, possibilities

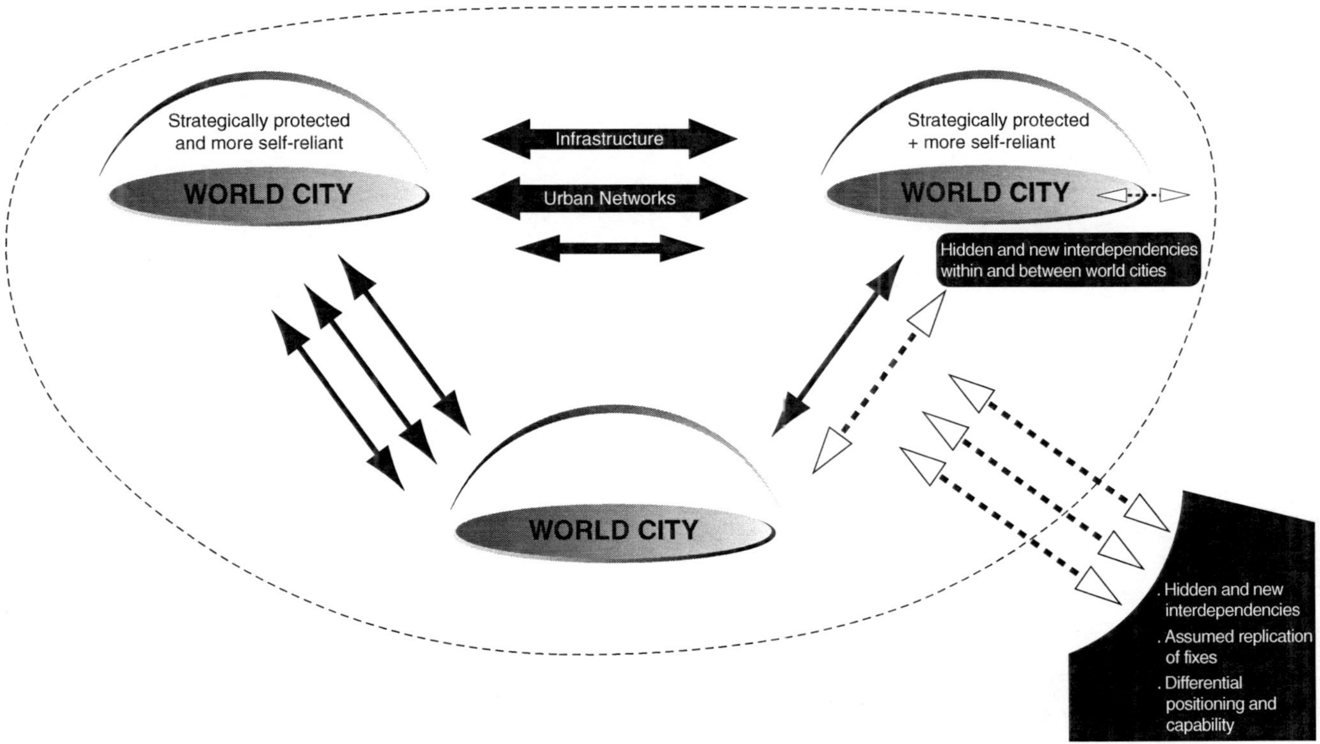

Figure 4.3 Archipelagos of glurbanized infrastructural agglomerations.

and necessity for reducing emissions in an economically beneficial manner. The growth of urban populations, their increasing levels of resource use and greenhouse gas emissions provides a context within which representatives of these leading cities are positioning them as the pioneers of city interventions and actions to respond to the challenges of climate change. Following the London summit in May 2007, leaders of municipal governments and international businesses and an expanded group from over 30 world cities convened in New York City for the second (what had become) C40 Large Cities Climate Summit. The cities promised a number of action points, which can be characterized in relation to three in particular: learning from and sharing 'best practice'; developing common procurement strategies; and devising standardized methods for measuring greenhouse gas emissions and responses to them.

Taking the first of these, by mobilizing and networking expert assistance the self-stated objective is to help cities develop and implement programmes that will lead to reduced energy use and lower greenhouse gas emissions in areas including building efficiency, cleaner transport, renewable energy production, waste management, and water and sanitation systems. A practical example is the development of a programme in retrofitting commercial and residential buildings in London, which has sought to position itself as 'leading the way in retrofitting buildings'.[12] Through its partners the C40 has, it is claimed, over 500 million square feet of buildings lined up for retrofitting programmes, through 250 projects in 29 different cities. Yet the process through which London constituted its retrofitting programme drew upon the experiences of Berlin and Copenhagen and then worked with the C40 to develop a model through which to 'guarantee' specific carbon dioxide (CO_2) savings within its own estate. The idea here being that the process constituted in London could be rolled out across the UK through a standardized procurement template so that developers and public agencies did not have to undertake a laborious process of constructing the contract themselves from scratch as the model was copied from Berlin. There is thus an assumed process that encompasses notions of transferability and replicability of 'best practice' from Berlin to London to other cities and places in the UK.

A second central function of the C40 is in developing common procurement strategies, where the coalescence of some of the world's largest and most powerful cities with some of the world's most influential businesses is being encouraged. It was proposed that the consortium would create partnerships with selected vendors with the aim of lower production and delivery costs and therefore lower prices of building materials, systems, and controls; traffic and street lighting; clean buses and waste disposal trucks; and waste-to-energy systems. Although procurement rules prevent cities simply combining all their purchases, the aspirations of the C40 were that – in representing 40 of the largest cities – there would be recognition of this collective power among suppliers and a subsequent discount.

The third area is the creation and deployment of common measurement tools so that individual cities can establish a baseline on their greenhouse gas emissions, track reductions and share best practice to inform mitigation and evaluation activities. Tangibly this means the development of standardized measurement tools. While recognizing that many of these different cities have produced climate change action plans they use different methodologies in doing so. Through working with Microsoft, the C40 has developed a standardized method to compare and contrast the setting and measurement of targets.

Networks of large cities as sites of 'practical action'

The intervention and actions of cities, through coalitions of interests, raises questions about their role. There is an emerging rhetoric from these groups in world cities that they have a responsibility to cut emissions due to issues such as their energy consumption and their vulnerability to the consequences of climate change. The role of cities as sites of 'practical action' in a context of national leadership in the response to climate change is also centrally promoted. The view of cities as practical actors sees a key networking role in building links between cities across the world.

There is an emphasis in the rhetoric of the C40 on sharing experience and mutual processes of learning which underpin long-term collaboration between leading cities coalescing around active and practical responses to the challenges of climate change. This emerging agenda focuses on where city leaders are seen to have degrees of control, in particular over greenhouse gas emissions in relation to their own facilities, buildings and fleets, where they can 'share information', develop 'best practice' and become climate change 'exemplars' to other cities. Nods to 'collaboration' are made on the basis of 'working together' to address common challenges facing major cities through a commitment to sharing best practice and through a pledge to pursue low-carbon procurement strategies that underpin the innovation and uptake of 'climate-friendly technologies'. It is on this basis in particular that 'partnerships' between large cities and corporate business are being promoted.

The emphasis of the C40 has been on the activities and responses of leading cities in respect of cutting greenhouse gases. In particular, from its earliest days as the C20, this has focused on the current levels of greenhouse gas emissions of cities, their targets for reduction and examples on how they are currently addressing these targets. This has addressed the activities of individual cities in terms of both mitigation and adaptation and their reduction. It has also emphasized the centrality of technological responses and key exemplar technology and infrastructure projects, which are promoted as low carbon technologies. Table 4.1 provides a flavour of a range of responses – in particular the targets, technologies, projects and costs – that C40 cities are developing in relation to transport. The representation of the C40, their

Table 4.1 C40 cities, targets and technologies

Transport

City	*Case study*
Barcelona	Bicing is changing transport modes in Barcelona
Bogotá	BRT system reduced travelling time 32%, reduced gas emissions 40% and reduced accidents 90%. CicloRuta is one of the most comprehensive cycling systems in the world
Copenhagen	City of Cyclists reduces approximately 90,000 tons of CO_2 emissions per year and has over 36% of the city's population cycling to work every day
Freiburg	Freiburg is an inspirational city powered by solar, where one-third of all journeys are by bike
Jakarta	A 12.9 km Bus Rapid Transport system built in just nine months at a cost of $2 million/km
London	London's congestion charge cuts CO_2 emissions by 16%
Mexico City	Mexico City replaced 3000 taxis with more fuel-efficient models
Paris	Velib – a new Paris love affair
Portland	Optimizing traffic signal timing significantly reduces the consumption of fuel. SmartTrips informs and incentivizes Portland residents about alternate forms of transportation
San Francisco	A world-leading low emissions transport system with zero-emission vehicles
Seoul	Seoul car-free days have reduced CO_2 emissions by 10% annually
Stockholm	Stockholm's clean vehicles are slashing 200,000 tons CO_2 annually Stockholm to introduce congestion charge – trial cut CO_2 by 14%, traffic by 25%

Source: C40.

aspirations and current practices raises a number of issues, to which we will turn now.

'Collaboration' in becoming climate change exemplars

Understanding these developments means addressing the ways in which the C40 represent their activities and roles, the stakeholders and relationships that inform the C40 alliance, and then opening to scrutiny their assumptions and claims. By privileging the role of cities as huge producers of carbon, major consumers of energy, as victims of climate change and contexts for action for addressing climate change this positions cities not as challenging the role of national governments but as responding to their leads through practical action.

Accelerating collaboration through fledgling relationships provides the context within which meaning is not pre-constituted but offers a focus for its particular constitution. Collaboration thus means that cities increasingly work together to tackle environmental challenges increasingly faced by major world cities – by which cities is meant, mayors or city leaders. Yet, the claim is that cities can learn from the experiences of each other and, thus territorial contexts are alluded to in populating the work of the C40 but in the process of doing so are symbolically disembedded from their territorial contexts. The C40 is underpinned by a 'commitment' to procurement of green technologies and related forms of green consumption, where a key tenet of addressing climate change is through technological innovation and its commercialization.

Working together and collaborating is extended to developing partnerships with corporate capital offering demonstrable – primarily this means tangible technological – ways to address climate change. The proactivity of cities and corporates is held together through a shared interest in becoming climate change exemplars who seek to demonstrate cuts to greenhouse gas emissions within a context of growing the 'bottom line'. This is an attempt to naturalize the view of action on climate change as being the domain of urban and corporate elites where world cities provide contexts for the demonstration of exemplar technologies and innovations of corporate capital. The linking together, through networks, provides economies of scale for the material manifestation of these initiatives but also a space for cultural negotiation and upscaling and buttressing of these particular territorial manifestations. Proactivity around climate change is framed within a view of leading world cities as exemplars and as providing 'practical solutions' to addressing climate change. They can learn from each other in sharing best practice, which remains an underdeveloped concept given its emphasis on targets, technologies and costs. It needs further population, which requires reflection on specific world city contexts from particular positions (e.g. mayors, city leaders and other associated officials), communication and translation to the shared space of the network of world cities.

C40: an exclusive network

Since the London summit in 2005, the C20 has expanded and has now become the C40, with 40 participating members and 17 affiliate members, with a partnership since August 2006 with the CCI (see Tables 4.2 and 4.3). The initial criterion for full membership is population size, with each C40 member having a city or urban area of 3 million people or more. In addition a category of affiliate membership was created to encompass cities smaller than this figure where it was felt that particularly innovative activities were taking place in respect of responding to climate change. So, for example, when Copenhagen announced a climate change action plan for carbon neutrality by 2025 the city was invited to join in order to learn about the process of working towards carbon neutrality and to asses what might be transferable to large cities. Yet, despite pressure for expansion, the C40 is a much more exclusive network than other networks of cities, such as ICLEI with its more than a thousand member cities, towns and regions.[13] The justification for this is on the basis of limiting size to make the network workable, where the role of C40 cities is to show leadership and act as a catalyst whose programmes can then be followed by any other

Table 4.2 C40 cities

Addis Ababa	Delhi NCT	Lima	Philadelphia
Athens	Dhaka	London	Rio de Janeiro
Bangkok	Hanoi	Los Angeles	Rome
Beijing	Houston	Madrid	São Paulo
Berlin	Hong Kong	Melbourne	Seoul
Bogotá	Istanbul	Mexico City	Shanghai
Buenos Aires	Jakarta	Moscow	Sydney
Cairo	Johannesburg	Mumbai	Tokyo
Caracas	Karachi	New York	Toronto
Chicago	Lagos	Paris	Warsaw

Source: C40.

city. Critically there are pressures for individual cities within the network to act in a transformative way with the sanction of being asked to leave the exclusive club if they do not.

Networks of practice

Coordination of the C40 is managed by a small secretariat based in London that oversees workshops – for example on transport and congestion in London in 2007, adaptation in Tokyo in 2008, ports in Rotterdam in 2008, airports in Los Angeles in 2008 – information exchange and website development. The practice of the C40 network and how its functions such as sharing best practice and learning manifest themselves involves a multiplicity of relationships, including variable relationships with other cities in the C40 network; relationships between C40 cities and business; C40 cities' strategic engagement with, and disengagement from, their national governments; and finally relationships between C40 cities and other cities outside the network.

Taking the first of these, the C40 works to coordinate senior officials within the mayor's office of each of the cities within the C40 groups. This they do in a number of ways, including through the two-yearly summits (London 2005; New York 2007; Seoul 2009) and through a range of workshops, as outlined above, where presentations of activities in different cities are made that illuminate ideas that are coming to the fore, progress that is being made and successes. Bilateral linkages also exist, for example in the case of London and Stockholm these are focused on decentralized energy,

Table 4.3 C40 affiliate cities

Amsterdam	Copenhagen	Portland	Stockholm
Austin	Curitiba	Rotterdam	Yokohama
Barcelona	Heidelberg	Salt Lake City	
Basel	Ho Chi Minh	San Francisco	
Changwon	New Orleans	Seattle	

Source: C40.

but these interactions are based on particular sets of personal and political relationships that are susceptible to a change in political administration.

The agenda to reconfigure urban infrastructures of the cities in the C40 and CCI is being led not only by a small number of urban elites but also by business elites. Between them they represent a highly select group of large cities and large corporates. Collaboration is the organizing principle of the C40 and those that are excluded are constructed as 'other' – that is as external to the emerging view of a small group of world cities and particular interests of corporate capital as of their necessarily central role in the active addressing of climate change. Notions of leaders and leadership are constituted around formal position and the strengths of symbolic allusion to that position – for instance it is 'obvious' why particular social interests should shape the agenda. In July 2006, for example, a Roundtable on Climate and Energy met in California and encompassed business interests and subnational and national government. Those involved in the discussions included Lord John Browne, BP; Charles O. Holliday Jr, DuPont; Sir Richard Branson, Virgin Group; Sergey Brin, Google; John Bryson, Edison International; Jacques Dubois, Swiss Re; Dan Hendrix, Interface; Michael Morris, AEP; James Murdoch, British Sky Broadcasting; Anthony Pratt, Pratt Industries/Visy; Tom King, PG&E; Jeff Swartz, Timberland; Tracy Wolstencroft, Goldman Sachs; Rick Lazio, JPMorgan Chase and others.[14] The central justification for this, according to Steve Howard, CEO of the Climate Group, is that: 'These companies represent one half trillion dollars in turnover, well more than 315,000 employees, and in excess of 207 million customers'.[15]

The active role of world cities in building relationships with each other and corporates through the C40 and CCI to share best practice and work together also involves a degree of reworking of relationships with national government. The administrative view of infrastructures, discussed in Chapter 1, was largely predicated on a national-level view of the subnational as relatively benign. The re-emergence of cities means that these national–subnational relationships are the locus for degrees of reconfiguration and bypass predicated on coalitions of world cities and corporates. The consequence of distancing cities as the contexts for action from national leadership is that it provides a context within which the acceleration of cooperation between cities can be justified as: the national leads and the city acts. Not only are cities bypassing national governments to collaborate with each other, but also they are doing so to collaborate directly with national governments and vice versa. This can be seen, for example, in the agreement struck between the Governor of California and the UK prime minister in 2006 to reduce greenhouse gases and promote low carbon technologies and also in an oil deal between the Mayor of London and the President of Venezuela in 2007.

This view not only excludes a variety of social and political interests within the context of the individual cities, but also ignores the contexts of cities outside the C40 network. It displays many resonances with the tenets of territorial urban growth machines and the narrow cadre of social interests often inhabiting these spheres. This exclusion is not portrayed as such, but rather as the elites of large cities and corporates acting independently of political affiliation and particular interests to address the threat to the world through climate change. This is underpinned by a politics, where 'key partners' collaborate in constructing their (narrowly defined) agenda of the 'measurable' results of their initiatives, where climate change is framed as an obvious economic benefit. This is to be achieved through reducing greenhouse gas emissions via a medium and long-term target approach which underpins investment in practices and technologies which translate from niche, experimental activity to the 'mainstream'. Collaboration underpins the basis through which the issue of climate change is framed as a problem in relation to large cities. It becomes the organizing principle around which city and corporate leaders coalesce and it provides a space in which the different territorial and organizational positions frame the 'natural' role of large cities and corporate business in addressing the problems of climate change.

Conclusion

A new agenda is emerging around the role of cities and their responses to an era of resource constraint and climate change. World cities and leading corporates are developing networks and coalitions which are defining the terms and shape of responses to climate change and resource constraint. In the context of world cities they are developing 'fixes' that appear to be prioritizing two scales of infrastructure connection: relative dis-engagement from national and regional infrastructure and the reprioritizing of city-based enclosed resources while at the same time seeking to ensure continued intra- and interurban connections through new urban agglomerations of socio-technologies. Such responses are wrapped within a wider issue of ensuring strategic protection and seeking to guarantee the longer reproduction of the city.

At the same time a critical issue is the extent to which strategic protection involves the relocation, the fortification of, the retrofitting of or the abandonment of existing systems, parts of the city and populations. In doing this it raises two critical issues. The first of these is in the processes through which strategic protection becomes mobilized. Who is protecting what? And, by deduction, what and who remains exposed?

Attempts to promote self-sufficiency in relation to world cities are predicated on the rescaling of resource flows to underpin new forms of growth in relation to particular places and to do this through organizing principles such as carbon neutrality, zero waste and energy independence. Economy and ecology become increasingly strategically interrelated. Coalitions of interests, that often include urban elites, corporate capital, architects and utilities are engaging in the deliberate encouragement of the visibility of these fixes to demonstrate technologies, techniques and approaches that are promoted as replicable to other contexts. That is to say that techniques and technologies become products that are exemplified in relation to the world's most powerful cities through showcasing and experimentation.

All this seeks to make obvious a particular geographical imaginary from a range of potential options. This geographic imaginary needs to be understood not solely in relation to the bounded context it seeks and claims to produce but also in terms of the potential new interdependencies that underpin reconfigured resource flows. So 'within' the circulation of resource flows there are attempts to create 'neutral' processes. Yet, it may be that often elements of the production and disposal of resource flows remain outside of the self-sufficient loop. One can think here of the case of new biofuels, hydrogen buses, solar panels, battery cars and the technologies of their production. Furthermore, there is the issue of the repatriation of profits made by corporate capital to their headquarter locations. In this respect the issue of the extent to which decoupling or withdrawal from wider systems of provision or the extent to which the city can be closed is raised.

The 'inevitability' of coalitions of world city and business elites is also couched in the depoliticization of their weighty political leverage through the mobilization of their assumed replicability and notions of collaboration and workability. There is little sense that the fight against climate change and resource security could be other than the solutions afforded by the huge efforts of these coalitions to couch the fight in terms of large-scale investment and the possibilities that technologies open up for developing new markets opportunities and 'bottom line' in highly visible leading city contexts, with their proximities to the architectures of world cities of media, political decision-makers, tourist hubs and visibility generally. Not only are these world cities positioning themselves and being positioned by corporates and some national governments as the exemplars of climate change action, but also they are pushing the message that they are cultivating best practice and (either implicitly or explicitly) that there is little alternative to this logic and that their actions can galvanize other cities through some subsequent transferability of best practice.

An emerging agenda such as this one raises many new research and policy challenges. There is also a significant politics to its production and a series of implications and consequences that follow from it. In Chapter 5 we critically consider the consequences of secure urbanism and resilient infrastructure.

PART III

Implications, Limits and Alternatives

5

The 'metropolitanization' of ecological resources

Introduction

In this chapter we focus on the implications of world city responses through secure urbanism and resilient infrastructure (SURI). Our concern is to critically explore whether SURI signals a shift to a more socially and ecological 'progressive' style of urbanism or alternatively whether this movement is best understood as the action of narrow coalitions of social interests, through large world cities attempting to ensure continued economic and spatial reproduction by securing privileged access to network resources. Consequently, we ask whether SURI is really about building bounded security at a metropolitan scale – thereby attempting to guarantee the continued availability of networked infrastructures within and between large world cities to ensure their continued reproduction – or is it about building collective planetary security by developing a more ecologically efficient and modernized style of urbanism that can be rolled out globally to guarantee fair access to ecological resources against a context of global ecological change.

Many aspects of SURI resonate particularly strongly with the type of environmental responses that the environmental movement has been working for several decades to put onto the urban agenda. For instance, the desire to create more ecologically efficient environments, such as green or autonomous buildings and the rapid roll-out of these technological solutions to create more 'self-reliant' cities, has largely been seen as socially progressive and ecologically desirable. Given this particular context we realize that it may be seen as problematic to ask difficult questions about the value and consequences of this emerging logic of SURI which may accelerate – due to the involvement of the corporate sector and urban governments – the development of apparently more sustainable and eco-efficient forms of urban development. This emerging form of self-reliant urbanism is seemingly almost universally supported by coalitions of social interests including corporate and environmental interests; it is assumed to be progressive while creating new economic opportunities; it is intended that eco-cities should be widely replicated and it is hoped that the fixes developed in world cities should be rolled out in ordinary cities. A coalition of social interests has emerged around these responses that is largely uncritical and seeks to promote the roll-out of SURI across multiple urban contexts.

But we are uneasy with the almost unrelentingly positive treatment of the eco-cities discourse. Our concern is that SURI is not being subjected to enough sustained critical scrutiny in order to at least test the assumptions and claims being made for the movement and its implications. We feel that there are important reasons why we need to start asking more analytical questions about the movement and try to uncover the tensions and contradictions in order to develop a wider politics of possibilities about different urban futures.

Urban history tells us that we need to subject architectural and urban planning visions of new ecological utopias to more critical scrutiny rather than simply accept the claims of the proponents of new styles of urbanism. Critical analyses of the new towns movement in the UK, ecological architecture after the oil crisis of 1973 and the promotion of gated communities have all demonstrated that the claims made for utopian visions of future cities invariably fail to live up to the claims of the social visions and technological expectations made for them by their proponents. This means that we should subject existing utopian visions to critical scrutiny and ask difficult questions about the implications of SURI now in order to better understand the differences between the claims made for them and the socio-material consequences of their translation into action. In particular we have stressed the need to ask whose social visions do these represent? Which social interests are included and excluded from the development of such visions? How do processes of restructuring existing social relations and eco-technical interdependencies unfold in relation to cities? And what socio-material consequences do they have in redistributing the benefits and costs of urbanization?

The double threats of ecological and economic crisis have helped to accelerate the naturalization of what are constructed as inevitable and positive responses by coalitions of interests in and through large world cities to pressing states of emergency. This tendency means that difficult questions about the politics of the responses – who they benefit, who is excluded – are hidden and not addressed because of the seemingly cross-cutting nature of the urban, environmental and corporate coalitions supporting SURI. The depoliticization of debates about the future configuration of cities' socio-technical networks and the new networked ecologies they represent is extremely dangerous. As soon as coalitions of interests talk about the inevitability of particular responses and the critical issue of accelerating these in other urban contexts – without any understanding of their consequences and the type of cities that we will be creating – this means that we actually need more critical analysis not less.

In subjecting these developments to scrutiny our concern is not to dismiss these responses out of hand, instead we want to ensure that wider societal learning takes place to really understand what may be missing but also what might be relevant and transferable. While we will ask some difficult questions about what we consider to be utopian visions that are not particularly well connected to a politics of the present, our concern is to enlarge the range of possibilities, to prevent overlying technological or productionist solutions that will be resisted and may not work. We want to find a way, through critical analysis, of seeing how we can learn about the limits and consequences of autonomy to develop fairer, more equitable, more robust and socially informed trajectories of eco-urban development. Consequently then we ask what lessons can be learned about a different style of urban development and other possibilities and alternatives by focusing on a critical analysis of the claims made for urban ecological security. In this chapter we raise six sets of issues in response to thinking about the consequences of SURI.

Producing 'new fixes': the naturalization of world city responses

World cities, through coalitions of interests, are attempting to develop new infrastructure fixes that – in aspiration – seek to reshape relationships between the city, infrastructure networks and critical resource flows in order to fulfil their economic growth ambitions. Central to this are the development of strategies that aim to ensure there is sufficient resource to supply cities' economies under conditions of climate change, resource constraint and the potential introduction of carbon reduction targets and carbon pricing strategies. Critically, new and opportunistic coalitions of social interests at and within the metropolitan scale – city governments, private and corporate interests, in some cases environmental groups – are being enrolled into these processes and have become active in shaping debates about the development of new configurations of city, infrastructure and resource flows. Increasingly it is these coalitions of social interests that are working within and/or through the world's largest and most powerful cities such as London and New York, cities of the C40 that are developing a more strategic orientation towards critical questions about their future resource requirements and infrastructural implications. A new dimension of the competitive positioning of cities is their ability to internalize, bound and control resource requirements as they seek to ensure the security of their systems of collective provision.

Certainly in the case of the large world cities, mayors and government structures at the metropolitan scale appear to be taking much more strategic interest in questions around the long-term ecological security of their cities. This is evidenced by the strategies of, for example, London, New York and San Francisco. Key to this is the critical involvement of private and corporate interests and environmental groups in the development of strategies and proposals for systemic change in the form of partnerships, demonstrations and test-beds. World cities and these economic-ecological coalitions are clearly positioning themselves as being the obvious actors and places to address the 'threats' of resource constraint and climate change. This we can see as a particular political mobilization of a rhetoric of depoliticization; the result being a particular framing of the agenda as obvious, where these partners collaborate in constructing their (narrowly defined) agenda of the measurable results of their initiatives where resource security and climate change are framed as economic benefits. SURI provides a strategic framework for organizing all manner of individual city targets around resource constraint and climate change.

The 'preparatory' metropolis: interdependencies, time and eco-knowledge

Rather than seeing ecological constraint as a limiting factor in cities' future growth ambitions, the new logic is to systemically anticipate and strategically prepare networked infrastructures for a period of ecological constraint. Consequently ecological issues become strategic within cities' future growth strategies as the ability to continue to grow is intertwined with the ability of a city to guarantee the ecological resources necessary to support economic growth. Longer-term questions about the availability, reliability, control, security and costs of ecological resource become central to long-term strategic responses at the scale of the city.

The emergence of the preparatory metropolis is linked to a major turn in how we understand the interrelationships between cities and their wider networked ecologies. This is partly a result of incidents such as 9/11, Severe Acute Respiratory Syndrome (SARS), heatwaves, oil price rises and so on. Taken together these incidents consistently demonstrate that the apparent stability of the technological, social and environmental relations that ensure that networked infrastructures work effectively is more unstable and fragile than previously understood. At the same time interdependencies across time and space are much more tightly coupled – if variably so.

Ever more tightly coupled infrastructure networks intensify time-space compression (Harvey 1990) to the extent that crises in one distant context actually generate serious problems in other urban contexts. Consequently the distinctions between rural and urban become more problematic as, for example, disease can pass rapidly through food systems and international aviation networks, and between the cities of rural China and the developed North. The increase in the frequency, scale, severity and impact of crises that range from the perceived threat of terrorist activity to concern about the implications of climate change on climate, energy, water and food systems has started to reshape debates about emergency responses. This is manifest in a shift away from more ad hoc emergency responses – the politics of urgency – to a new type of response in which preparedness is institutionalized based on a strategy of building resilience. Increasingly the challenge in seeking to build resilience lies in strategies that can accommodate the shifting relations between the technological, social and ecological that are adaptable to the contingencies of specific crises but that also maintain greater degrees of continuity under the new pressures.

World cities, through coalitions of interests, are developing a preparatory strategic orientation to resource issues and are increasingly using much longer-term timescales in the formulation of their research, planning and policy priorities. These cities seek to anticipate and prepare for long-run ecological resource changes and do so by attempting to effect complex socio-technical change in the organization of fixed and relatively obdurate infrastructure systems strategies. As world cities start to grapple with the development of urbanism in the anthropogenic period, questions around long-term climate change mean that new timescales are being incorporated into planning processes as is a concern with understanding the interconnections between ecological, economic and technological changes. What this means is that 'normal' planning processes that focus on relatively long five- or ten-year periods for understanding economic change and planning pressures and priorities now have to be overlaid by much longer-term timescales that deal with systemic changes in weather patterns, sea level, climate and resource availability (energy food and water). In research terms cities are being pushed into, at least, framing economic and spatial issues within extended timeframes of climate science, climate change and ecological changes in energy, food and water systems. These strategies are increasingly characterized by much longer-term planning horizons. Climate change and planning strategies are being developed up until the end of the twenty-first century while infrastructure strategies are commonly linked to timescales that reach to 2025 and 2050.

Part of the reason why these long timescales are necessary is the slow moving nature of change in infrastructure systems, which may have lifetimes of 40 to 100 years. Consequently in making decisions about current investments in transport, water and energy infrastructure, coalitions of interests are having to do so against planning timescales that grapple with the urban implications of global climate change in their own particular city. Obviously this is being done against a background of significant uncertainties and risks in understanding the implications locally, but representatives of cities are being pushed into trying to systemically prepare by understanding the changed conditions in which they will attempt to ensure the long-term reproduction of the city. Consequently, there is clearly a political element to these longer-term timescales.

Critical here are the ways in which large cities are the locus for attempts to try to generate understanding of the implications of claimed change and resource constraint in their own particular context as part of strategies to try to capture national resources to ensure their longer-term strategic protection. Coalitions of interests in world cities are actively competing within their own national context and internationally to seek to guarantee that they are subject to critical infrastructure protection, flood protection and preparation to ensure that they are among the best equipped to deal with climate change. The significance of this is in understanding whether this longer-term thinking is an appeal to the precautionary principle, to recognize that longer-term costs of consumption have implications of geological significance, or whether they are an attempt to gain first mover status and capture priority within national and regional plans and priorities.

New forms of knowledge, expertise and social interests are being brought into the strategy development process at the level of the city to attempt to develop an understanding of issues where city authorities may lack formal responsibilities and or any experience. This places a premium on the abilities of city authorities to engage and enrol new scientific and technical capability with respect to climate change, energy security, resource security, technological change and in understanding the implications and options for the economy and place-based priorities. New relationships of knowledge production and exchange between cities and consultants, universities and utilities are becoming increasingly critical in generating local understanding of pressures and the implications for urbanism and infrastructure.

Increasingly in world cities we are seeing a prioritisation of the development of place-based understanding of the implications of climate change and resource constraint on urban climate, resource flows, carbon strategies and so on. It is in the major world cities that sustained attempts are being made to develop a new scale of climate science with a central role in this for climate modelling, climate science and resource understanding of ecologies of cities. In ordinary cities or cities of the global South such understanding is often either absent or requires charitable trusts to fund relevant research. This is a very particular geography that is emerging, where the most well developed understanding of the impact of climate change is focused on a few large cities. Critically coalitions of interests are trying to develop expertise to position them to benefit from first mover status and ensure strategic understanding.

The development of new knowledge about urban resource flows increasingly subjects the ecologies of cities to monitoring, modelling and scrutiny. Ecology itself becomes subject to specialist calculation, predictive modelling and monitoring. The very ecology of the city becomes laid bare as increasing understanding of flows of resources is built and new spaces and ways of representing climate change, graphs, energy flows, water consumption and the carbon density of different sectors are found. Cities' ecological flows become selectively made visible and hidden. But critically what is often missing from these understandings of network ecologies are the politics of their production and the socio-technical organization of resource flows.

A key issue, then, is how these forms of knowledge and expertise are designed to work. One distinction is between whether coalitions of knowledge and expertise are asking critical questions about how we may imagine a redistribution of resources based on fair shares, equity or some notion of environmental limits or whether, such coalitions, instead, signal a new form of urban eco-managerialism designed to capture resources for particular places so that they can ensure their continued reproduction. It is no surprise that in these spaces new forms of knowledge providers – both academic and commercial – are emerging offering specialist expertise to understand resource metabolisms in relation to the city, to generate specialist intelligence and understanding of threats to cities and corporates and in turn to provide knowledge of how they can build resilience and security for these critical resource flows. However, we need to understand that such knowledge is often limited in terms of the breadth of the constituencies that produce it and, as such, there is a significant shift to think about strategic protection and security in increasingly narrow conceptions of bounded security as science focuses on knowledge that protects resource flows that are critical to production–consumption systems.

'Metropolitanization': relocalization and glurbanization of networked ecologies

The critical consequence of world cities' responses are characterized by an increasing metropolitanization of networked ecologies. The dominant strategic response, as we saw in Chapter 4, is exemplified by three interrelated strategies of reconfiguring socio-technical infrastructures in order to secure strategic protection from climate change and resource constraint for the metropolis; to attempt to build autonomy in the provision of key ecological resources such as energy, water, food and mobility; and, to construct new urban agglomerations of infrastructure both within and between world cities. Consequently the metropolitanization of networked ecologies refers to the dual development of a systemic and strategic orientation of major world cities towards ecological resource through two forms of rescaling of their socio-technical infrastructures.

Strategic relocalization infrastructure and resource flows

Strategic relocalization is based on a simultaneous process of the rescaling of socio-technical infrastructures 'upwards' from green buildings, local experiments and test-beds to more strategic and systemic transitions at the metropolitan level and 'downwards' through active withdrawal from external reliance on existing regional and national infrastructures. Strategies for the development of decentralized energy, water and food systems are increasingly upscaled from individual buildings or technologies. These niche activities are developed into an aspiration for systemic transition in the social and technical organization of decentralized infrastructure at the scale of the metropolis, as new coalitions are constructed claiming to speak on behalf of cities' attempts to build greater autonomy and reduce reliance on external infrastructures and resources. Because the aspiration of greater autonomy has to be actively constructed the process of building territorial self-reliance is paralleled by a political strategy of withdrawal from and bypassing of existing infrastructure networks at the regional, national and international scale as metropolitan scale infrastructures are reprioritized.

Relocalization is best understood as a process through which world cities, ecological resources and socio-technical infrastructures are increasingly reincorporated and re-enclosed within or proximate to the metropolitan boundary. Such strategies seek to develop decentralized technologies, local food production, closed-loop systems, and more circular urban metabolisms while simultaneously aiming to reduce cities' reliance on more distant and uncertain resource flows and infrastructure. This is likely to create new infrastructural controversies and politics as place-based territorial interests start to use (financial, institutional and knowledge) resources, policies and investments to try to secure the development of particular infrastructural configurations. These are likely to be manifest as an intensified place-based politics around particular infrastructural fixes that seek to accelerate relocalized options and selectively reduce reliance strategically on regional, national and international networks. This may be in tension with the strategies of established infrastructure providers who have wide network remits or are seeking to develop solutions at enlarged scales.

Strategic glurbanization and archipelagos of networked ecologies

At the same time the glurbanization of networked ecologies produces two forms of rescaling as selective networks of world cities work collaboratively to upscale the development of new infrastructure fixes within and between cities and downscale the fixes developed in world cities into other cities within the national urban hierarchies. World cities are working collaboratively to share best practices, knowledge and expertise in the development of strategic protection and relocalization of resources as they seek to match the performance of the 'best' cities in terms of recycling, energy independence, water savings and so on.

At the same time recognition of the limits of autonomy especially in relation to fuel supplies for transportation means world cities are collaborating in the development of global agglomerations of new infrastructures such as biofuels and hydrogen as they seek to develop new socio-technical systems that can secure such cities' future mobility strategies. Networks of large world cities are occupying a key role in the development of such infrastructure fixes and constitute a new scale for the development of socio-technical configurations. Critically world cities are being positioned as national and international exemplars for the development of new fixes that can be cascaded down on to other cities in the urban hierarchy. Cities like London and San Francisco are acknowledged by national government as being centres of innovation and sources for new infrastructure configurations that can be replicated in other contexts. Consequently there appears to be a parallel process of world cities up-scaling their attempts to develop new infrastructural configurations and a subsequent down-scaling through the national urban hierarchy of such fixes to other less dominant cities.

New global agglomerations also refer to the complex networks of infrastructures that connect up the global system of world cities. The global infrastructures of airline systems, aviation freight logistics, commercial seaborne freight logistical systems and telecommunications networks play a critical role in connecting large world cities. Container ports, international airports, logistics and distribution centres while not necessarily under the direct control of large world cities are usually located in close proximity and constitute a critical networked ecology strongly shaped by world cities but also reinforcing of world cities' roles as critical command centres. As we have seen, the ecological costs associated with these infrastructures are extremely significant both in terms of energy consumption and carbon emissions. Increasingly representatives of world cities have become active in finding ways of ensuring that these infrastructure linkages, which are seen as critical to the long-term economic competitiveness of such cities and future development scenarios, are not restricted in their growth or development.

Consequently an infrastructural politics appears to be developing around these networked infrastructures whose maintenance and security is critical to world cities. For instance a new politics around carbon measures is emerging which is selective in its treatment of aviation-related emissions. Thus strategies of displacing responsibility of these emissions 'outside' of the city could be explored. Likewise other strategies focus on the emissions directly associated with a particular bounded space. So, for instance with shipping, large polluting ocean-going container ships are required to use alternative fuels or energy sources – but only when they are in ports! Outside the boundary of the port they can use polluting technologies.

Critically then the treatment of large-scale global infrastructure that connects world cities becomes a particular problem and strategies are likely to be adopted – displacement effects, accounting procedures, territorial space shifting and so on – to not find ways of maintaining networks outside of ecological constraints. More recently the UK government announced that there will be no restrictions on aviation growth and that carbon reductions will have to be made up from other sectors. World cities that particularly benefit from the connectivity of global infrastructure do not want to be territorially responsible for the networked ecologies that lie outside the city. Although world cities may encourage alternative options – such as hydrogen biofuels and electric vehicles – the problems with all these solutions is that the ecological costs are produced outside the city.

Consequently, we are seeing a new networked politics, where selected world cities are represented as developing new infrastructural fixes, together and collectively, but are then cascaded down through the urban system. Primacy is given to maintaining global networked infrastructures and attempts to displace costs and not to internalize them within the enlarged metropolitan area. Yet at the same time the concern is to maintain high levels of interconnectivity between world cities and to ensure their economic positioning.

'Eco-competitive' cities: transcendence, markets and replication

Critical to developing strategies to ensure future resource requirements at the level of world cities is the incorporation of resource endowments and the ability to overcome ecological constraints into the economic and social competition between cities. At one level this is another performance indicator in the competition between places, but at another it becomes a more profound indicator of the ability of cities, through those coalitions that stake a claim on the city, to (at least in aspiration) provide the conditions that can 'guarantee' their social, economic and material reproduction in a period of resource constraint. What has shifted here is a move from the post-9/11 agenda of critical infrastructure protection from terrorism or the consequences of environmental damage to a position where the city can guarantee its material resource state against a background of resource constraint and competition. In a very real sense a new dimension of the competitive positioning of cities is an ability to internalize, bound and control resource endowment, supply, consumption and production.

Linked to the aspiration of greater ecological and infrastructural self-reliance is the claim that ecological urbanism can develop cities in almost any urban context overcoming both local environmental limits and anticipating a period of climate change and wider resource constraint. No context or environmental constraint is seen as a limit to the advance of ecological urbanism. This is a new style of urbanism that provides the technological solutions and market frameworks not only to overcome what we have conventionally understood as limits (water stress, limited infrastructure etc.), but also to prepare and anticipate a period of climate change (sea level rise, weather changes and so on) and ensure and organize continued reproduction under a period of resource constraint (mediated through energy security, food security, carbon neutrality and so on). Given impending eco-emergencies ecological urbanism will (attempt) to provide a guarantee that it can transcend any ecological limit.

Such strategies are seen as providing a new focus for economic accumulation in their own right, where pricing technologies, carbon trading, decentralized technologies, new fuels, and the physical fabric of the city are all presented as new opportunities for the development of ecological and economic services. It is in this light in particular that the active integration of ecological, infrastructural and economic strategies at the level of the world city are being promoted – by coalitions of interests that include city representatives, national governments, corporate and environmental groups – as a means to ensure the replication of the broader landscapes of SURI in other national and international contexts. In this way, world cities are being positioned as national and even international exemplars; that is key sites for experimentation, demonstration and development of new infrastructural technologies that are widely seen as having relevance and application in other urban contexts domestically and internationally. National governments often have a key role in developing these emblematic configurations both within and across different contexts. Critical to such strategies are the development of emblematic spaces for the piloting and experimentation of technical fixes within wider regeneration strategies in world cities.

Ecological enclaves: the limits of 'bounded autonomy' and 'archipelagos of global infrastructure'

It is clear that SURI is emerging as an emblematic technical fix, that is developed in particular world city contexts, that are viewed nationally and internationally as producers of innovations and that these fixes are seen as being replicable in other contexts. Paradoxically, despite such claims, SURI is largely a set of social expectations and visions that have yet to be translated into widespread material action through social and political networks. The production of such configurations has not been significantly tested in practice and consequently we are concerned that there is so much interest in replication when the feasibility of implementing SURI is not systemically understood. There are further substantive issues that already give serious cause for concern in strategies that seek to replicate SURI in a range of differing contexts.

The emphasis on the construction of 'self-reliant' cities through decentralized technologies – in the form of renewable energy, biodiesel, grey-water recycling, energy from waste, and the wider metaphors of closed loops and circular urban metabolisms – actually misses the point that infrastructures and cities are never truly bounded or autonomous spaces. It is a relative form of autonomy that reconfigures relations rather than provide total independence for the city that can, at most, be realistically built. While decentralized water production and reuse and energy production may decrease reliance on centralized networks they still imply interdependencies. How such relations are renegotiated and what implications these have are not yet well understood.

At the same time metaphors of circular metabolism are not often entirely accurate in their representation of resource flows which are frequently dependent on 'external' relations. An 'internal' focus on the development of biodiesel and hydrogen as new mobility fuels often tends to concentrate on consumption and ignore wider questions of resource production and the new sets of relations and dependencies set up in the low-carbon, post-peak oil city. It might be more accurate to think about the rescaling and reconfiguring of social and technical relations through the reorientation of resource flows and infrastructure rather than the construction of independence, closed loops and circular metabolisms.

Resistance to withdrawal and 'bypass'

There are a set of outstanding issues about the degree of social and political resistance to strategies designed to construct SURI and to its replication in other urban contexts. Many strategies have a strong techno-economic and productionist bias rather than focusing on behavioural and social change and reduction in demand for key resources. Central to this is the extent to which within coalitions of interests in world cities, as the developers of SURI, have to overcome local and community resistance to new infrastructural fixes in order to ensure their implementation. London's publicly accessible hydrogen bus refuelling station, for example, only finally received planning assent following a public inquiry that decided the station should go ahead given its national strategic importance despite significant and sustained local opposition. There are also critical questions about the capacity and capability in relation to cities and regions' lower down the urban hierarchy to develop and implement the type of fixes piloted in world cities around decentralized technologies and wider issues about the desirability of technical fixes developed in one context and their assumed transferability to other contexts without a clear understanding of their relevance or applicability.

Exploring alternatives: uneven capacity, social innovation and competing transitions

The active role of world cities and corporate business in addressing resource security and climate change through SURI is predicated on securing the future and boosting the economic competitiveness of world cities; but this also ignores the contexts of ordinary cities and many of the cities of the global South. This is not to say that responses are not taking place in these different contexts, more that they are likely to be more piecemeal and episodic in their responses to the pressures created by global ecological change. The reconfiguration of infrastructures under SURI becomes closely intertwined with the aspirations of corporate and city leaders and the 'market opportunities' of resource constraint and climate change. There is little sense that responses to resource constraint and climate change could be other than the solutions afforded by the huge efforts of coalitions of interests to promote the debate in terms of large-scale investment and the 'possibilities' that technologies open up. This offers some sense as to why these responses are developing in certain ways in particular places. But there is also a critical need to understand alternatives and their relations with the dominant response of SURI.

Strategies being produced in world cities are characterized by increasingly novel approaches focused on the managed shaping of systemic change in their infrastructures. The critical feature of this is that coalitions of interests in cities are developing their own place-based visions of the style of reconfiguration envisaged in their networks as they attempt to shift from centralized to more decentralized provision, reduce reliance on external resources, or find substitute fuels for mobility systems. Increasingly such efforts are linked to 'low carbon' transitions in the organization of infrastructure. This implies a much wider, more broadly based and complex view of how cities desire – at least in aspiration – to effect systemic change in their infrastructures, even where they do not own networks or have weak regulatory capacity to shape infrastructure and resources. Cities are differentially positioned with respect to the capability to develop, design, implement and learn from these new infrastructure solutions and this relates to the types of knowledge they produce.

World cities are actively being positioned as 'producers' of new infrastructure solutions through their representation as emblematic sites of exemplary innovation. London in the UK and California in the United States are both positioned as exemplary sites for the development of new infrastructure solutions. In contrast, ordinary cities often have no explicit national status as a producer of fixes. On the contrary, the dominant assumption is that innovative solutions developed in world cities are then translated into guidance and practice which is then consumed by other regions. The new environmental solutions being developed in world cities are often designed to exceed – either in level of ambition or by bringing forward their implementation in time – existing national environmental standards for resource use or waste production. In contrast, the targets in ordinary cities are usually designed to achieve existing standards rather than then exceed them.

World cities have a governance model that can more easily create connections to national government. For example, both London and Thames Gateway in the UK have a central government team specially involved in providing higher level governmental capacity and coordination for the development and implementation of infrastructure strategies. In contrast, there is no central government team within the centre that has the same level of responsibility or capacity for steering major regeneration developments in ordinary cities. World cities are the locus for developing the capacity and capability to innovate in the development of new infrastructure solutions. New forms of knowledge, expertise and social interests are being brought into the strategy development process as city authorities attempt to develop an understanding of issues where they may lack formal responsibilities and/or any experience.

In contrast, there is no comparable systemic attempt to develop solutions at this scale involving multiple partners, stakeholders and publics in ordinary cities. While there has been relatively slow progress in translating these solutions into practice the scale and scope of the ambition in the world cities is significant. New development projects and sports events are being used to act as large-scale demonstrators of innovative solutions. This is in contrast to ordinary cities where the largest or most emblematic development projects, with their potential as exemplars for innovative infrastructure solutions, remain as disconnected niche activities rather than as part of systemic social visions. Not unconnected from this point is that while world cities seek to anticipate and prepare for the long-term ordinary cities generally utilize much shorter-term planning timescales.

Social innovation and alternative networked ecologies

We have mainly focused on SURI in world cities. In other contexts different responses are emerging as we saw, for example, when we reviewed the Transition Town initiatives. Such strategies imply a more collective approach to innovation around climate change and resource constraint not solely oriented around technical fixes but suggest a more socially and culturally driven approach to new solutions and configurations. Critically these are designed *in* context and cut across all aspects of urban life through promoting resource reduction rather than the productionist bias of SURI. Similarly, emanating from the United States, under the rubric of a relocalization strategy, a network has developed that draws together over 172 urban post-carbon groups worldwide. The network developed in response to the environmental, social, political and economic impacts of global over-reliance on cheap energy.[1] Such approaches imply a more critical view of our reliance on current levels and forms of energy and the implications this has produced. Evidently there would be significant worth in looking further at such alternatives and how they compare and contrast with the strategies involved in SURI. There would be value in contrasting the different logics in terms of the social interests, the solutions developed, the balance between productionist and demand solutions and the implications of such strategies. More widely there would be benefits in considering how other constructions of SURI could be based on other concepts such as mutual interdependence, relationality, trading and trade-offs, fair shares, and environmental justice.

The strategies developed in world cities and the assumption of their relevance to and replicability in other cities are powerful. Yet these approaches highlight limits in their assumed transferability to ordinary cities and how they are seen by national policymakers. There are definite limits to transferability, including the ways in which such solutions often ignore the importance of local context, history and circumstances, in particular the forms of local knowledge, expertise and capacity that are available to translate the propounded solution in locally 'relevant' ways. But, there are also useful lessons to be learned from the approaches in world cities including the importance of developing systemic, long-term approaches to transitions in urban infrastructure and the ways in which this is fundamental to the continued economic, ecological and material reproduction of cities.

A key issue is how systematic approaches to future infrastructure provision and utilization in existing cities and communities are developed, which recognize relationships with surrounding environments and geography. It is important to recognize that transitions in urban infrastructure do not start from a blank sheet but from existing systems of infrastructure organized (politically, technically, production–consumption, regulatory) in a particular way, and to ask how this can be built on and/or be adapted to address broadly constituted priorities and a shared vision of the many different social interests who will have and should have a view on the future of urban infrastructure. Questions around urban infrastructure then are not only about the technical but also about widening the framing of infrastructure to involve the concerns and also the potential offered by a wide variety of social interests including utilities, regulators, developers, residents, citizens, environmental groups, and business.

Conclusion

In this chapter we addressed the implications, limits of, and alternatives to, attempts to develop and 'implement' secure urbanism and resilient infrastructure. We assessed the key social interests involved in developing SURI and the naturalization of world city initiatives as exemplary and emblematic of urban responses. We then moved on to identify the issues involved in the development of preparatory urbanism, in particular the resilience turn, the emergence of new timescales and the role of eco-knowledge in understanding the implications of anthropogenic change on cities. In addition, we detailed the issues involved in the rescaling of cities that prioritize the development of networked ecologies that are more enclosed at the scale of the metropolis and the infrastructural connections between world cities while withdrawing from reliance on infrastructure networks at other scales. We also explored relations between ecological and economic issues in cities asking whether the SURI agenda represents a new form of ecologically transcendent urbanism that provides a context for the development of new markets for infrastructure networks and services. We moved on to ask what the limits of autonomy, bounded spaces, selected archipelagos and internalizing resource flows are. In particular, we examined the creation, through SURI, of new interdependencies and forms of displacement that are less ecological and socially progressive than claimed. Finally we addressed alternatives to the dominant logic of SURI by outlining other types of responses that emphasize different priorities.

SURI represents an attempt to develop exemplary and emblematic responses that it is claimed can be simply translated into other urban contexts. Instead we argue that such claims need to be subject to far greater critical scrutiny. In particular the claims that these responses are inevitable and positive should not be treated as self-evident or even desirable. Instead the social interests behind such solutions, the claims of transcendence, the politics of rescaling at a metropolitan and world city system level, the assumed synergy between economic and ecological priorities all need to be critically assessed and tested. Our concern is that these efforts to develop an ecological fix are in fact not about transcendence and planetary security, but about attempts to displace environmental costs and build ecological security at a metropolitan level and between world cities.

SURI represents a narrowing of the conventional sustainability discourse with the squeezing out of debates about public participation, the social organization of resources and the issues of equity and fairness in access to environmental resources in order to build collective planetary security. This involves a much tighter focus on corporate and governance responses, a re-focusing on the economic potential and commercial logic of such responses with a much narrowing focus on place-based and bounded metropolitan security. Overall there is evidence that is sufficiently convincing that we need to think carefully about the wider conceptual and normative implications of SURI. In particular we need to be careful to guard against the development of a more bounded and divisible form of urbanism that does not contribute to the building of collective planetary security. It is necessary to subject SURI to much greater critical conceptual and empirical analysis as the dominant strategic response by world cities to the pressures of climate change and resource.

6

Conclusions:

Producing urban ecological security

Introduction

In the conclusion to *World Cities and Climate Change* we revisit our initial aims in writing the book. Using the theoretical and conceptual approach that we have developed and the evidence we have presented we argue that we are witnessing the emergence of a concern to achieve 'urban ecological security' (UES) and that this represents a paradigm challenge to our conventional understanding of contemporary urbanism. Within this broader context, a particular response – that we have called SURI – is exemplified and prioritized through a new style of infrastructural development that represents an attempt to build divisible and bounded security within and between world cities. We are troubled by the narrow politics and the consequences of SURI. In this chapter, we build a framework through which to populate conceptions of a more socially just and collective urban environmental security which provides the possibilities to challenge the underlying notions of SURI.

The emergence of 'urban ecological security' as paradigm challenge

Urban ecological security is defined as the selective incorporation of ecological issues into wider strategic debates about the approaches and responses required by cities in order to attempt to ensure their continued social, economic and material reproduction in the context of global ecological change. UES is, as we have demonstrated, a consequence of the combined effects of a set of pressures – the effects of global urbanism in reshaping

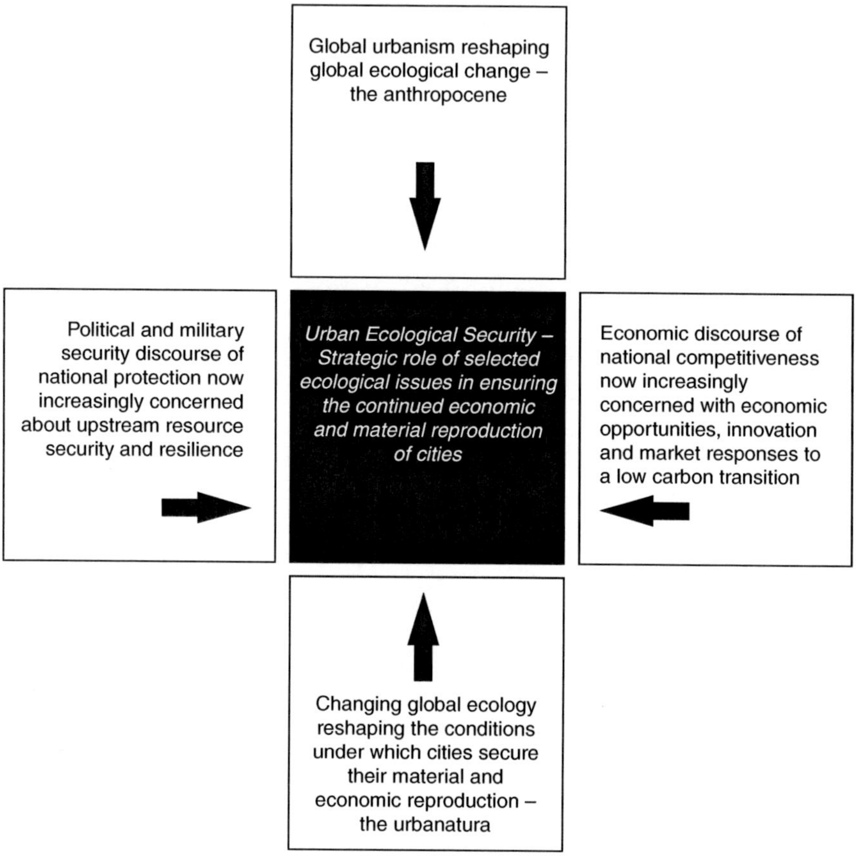

Figure 6.1 The emergence of UES: pressures, mediation and implications.

anthropogenic climate change; the extension of a political and military security discourse of national protection to encompass upstream resource security and 'natural disasters' in urban contexts; and an economic and innovation discourse of competitiveness and the ways in which it conditions responses at urban and regional scales. It is the organization and mediation of these pressures at an urban scale and the variable ways in which these pressures are experienced and responded to that reshapes the conditions under which the material and economic reproduction of cities is secured (see Figure 6.1).

UES becomes selectively strategic and critical to the continued economic and material reproduction of cities and regions under political security and economic competitiveness discourses. The ability of cities and regions – through coalitions of interests that constitute capability – to understand strategic threats to them through climate change and resource constraint and to develop systemic responses – the rebundling of network ecologies – becomes critical to their continued reproduction. We recognize that this is a selective and uneven process that does not apply in every place across all infrastructures or all cities.

Given the relational linkages between global urbanism, anthropogenic change and changes in the context for the ecological and material reproduction of cities there is now recognition that the societal processes of global urbanism are relationally reshaping the very ecological conditions under which contemporary urbanism has to reproduce itself. Yet the mediation of these global ecological changes – through the logics of military and political securities of infrastructure and ecology and a competitiveness and regulatory view of the economic opportunities of carbon control and low carbon transitions – means that cities are under increasing signals and targets to prioritize their own political security and economic competitiveness in understanding and responding to these pressures.

Consequently, the positioning of cities as critical sites for the intersection of these global ecological pressures and the combined logics of a security and competitiveness discourse *selectively* makes ecology and infrastructure strategic to future material and ecological reproduction. The urban resilience turn reflects this need to proactively prepare cities and infrastructure for climate change and resource constraint in order to protect and enhance place-competitive positioning relative to other urban centres. This also implies significant and systemic change in the socio-technical configuration of built environments, infrastructure and resource flows as cities seek to rebundle together new network ecologies. Reflecting on the wider implications of this turn raises critical issues.

The critical issue for us is that new pressures are eliciting responses that seek to appropriate the selective 'metropolitanization' of infrastructure networks to secure the material and ecological resources necessary for the reproduction of certain cities and populations within them under conditions of UES. UES both creates the conditions to which urban coalitions of interests respond and also informs the strategies which are produced to achieve an ecologically secure urbanism. Conventional responses to resource issues have tended to focus on constantly extending the reach of cities through more complex socio-technical networks and logistical systems that draw in ever more distant energy, water, food and material resources. Yet concern about urban ecological security challenges such traditional strategies. This presents nothing less than a paradigm challenge to urban studies which requires a shift in thinking about what is required conceptually and empirically in researching and responding to these issues.

It is important in seeking to understand urban reproduction to acknowledge the importance of urban (political) ecology. Our argument is underpinned by the view that economic, ecological, social, material and technological reproduction are increasingly becoming inseparable in debates around the future direction and understanding of urbanism. Securing urban economic and spatial reproduction cannot be disconnected from questions of ecological, social, material and technological reproduction and the social and political tensions that arise from them at multiple scales. Addressing this, conceptually and empirically, informs understanding issues of power, politics and possibility in the framing of and responses to UES and transcendental urbanism. We are concerned with how the complex dynamics of three interrelated sets of issues – economic reproduction, resource constraint and climate change concerns – becomes manifest and is translated into a dominant agenda.

In the production of UES cities are frequently characterized in three particular ways: as large 'consumers' of resources and as 'producers' of carbon; as potential 'victims' of the consequences of climate change; and as potentially innovative contexts of response to these challenges and issues. Coalitions of interests attempt to shape, mobilize and translate these discourses in relation to particular cities. This is addressed through entangled and multiple sets of relationships, across different scales of action, from different positions, where places are seen variably, and privileged and marginalized accordingly, by national states. There is a differential positioning of cities in their capacities and capabilities to act both in shaping and responding to UES. We have sought to make the entangled sets of multiple and multiscale social, economic, ecological and technological relationships and the ways in which they produce differential material and discursive manifestations of UES more apparent. In short, it allows us to develop more clarity around not only the social interests mobilizing UES and responses to it – and the exclusions from such processes – but also the differential capacity and capability available and developed to do so.

There are issues and limits to the new approach that this book addresses. For instance we need to recognize that many of the representational visions that have been produced for network transitions and rebundled network ecologies have not or have only partially been translated into tangible activities. Additionally we acknowledge that much of the power of these visions is in their assumed replicability in other contexts. We do not have to scratch far beneath the surface of UES to encounter claims of the 'transcendental' nature of this form of urbanism; claims that both the pressures of UES and the strategic responses to it have a universal relevance and transferability. What are ignored in these assumptions are the particular histories, capacities, capabilities and positions in national governance frameworks of those cities that are assumed to be able to replicate such fixes. These discourses mask that what is being discussed is a form of relative autonomy constituted not by self-sufficiency in resources and relationships but by new sets of relationships, interdependencies, displacements and 'costs' and, consequently, winners and losers. We, therefore, critically question the extent to which these strategies are really preparatory and anticipatory of global ecological constraints or whether they actually defer environmental costs in time or displace them in space as cities attempt to build the 'divisible securitization' of resources.

In the face of the naturalization of this attempt to secure and metropolitanize ecological resources, there is a critical need to understand the emergence of this dominant new logic. Systemic and purposive transitions in infrastructure are part of anticipatory long-term strategies to 'prepare' the metropolis for future economic, ecological and material reproduction. A central commitment embodied in this book has been not only to make visible questions of power and politics and their consequences but also to offer the potential for things to be 'other' and to offer visibility to alternatives where it is deemed to be obvious and inevitable. In line with our approach, we have aimed to demonstrate this theoretically, conceptually and empirically and, in doing so, provide the potential for ordinary cities to engage in more purposive, contextually sensitive transitions that inform their economic, ecological, social and material reproduction.

SURI as 'bounded and divisible' security for world cities

UES reshapes the context within which all cities attempt to ensure their continued reproduction. As we have made clear, above, there are significant differences between cities – in their positioning both within national contexts and between cities globally – in their capacity and capability to respond strategically to UES. While we have developed an overarching typology of the types of responses being developed by different cities as they attempt to produce new, more secure, networked ecologies we have been particularly interested in the exemplary and emblematic solutions developed by world cities. In doing this we critically articulated the importance of conceptualizing and understanding the ways in which space is and can be politically mobilized.

In particular we outlined how the differential mobilization of understandings of space – representational, relational, relative – underpins the political, ideological and institutional production of cities, infrastructure and resource flows. The deeply political and uneven organization of these relationships is driven by the interplay of representations of place through space as contained and fixed. These are relationally produced through particular institutions and social interests that prioritize and privilege particular places and populations, notably the interests of world cities. Specifically in relation to world cities we have argued that coalitions of interests are engaged in the development of three interrelated sets of strategic response which we have characterized as attempts to build secure urbanism and resilient infrastructure (SURI).

The first strategy concerns the degree to which world cities are strategically protected in terms of their preparedness for adaptation – dealing with heat and enhancing flood protection infrastructure. The second strategy addresses the way that world cities' resource bases are more strategically self-reliant, which is how such cities can attempt to guarantee sufficient access to key energy resources and internal mobility through relocalization, withdrawal and bypass of incumbent infrastructures. The third strategy deals with the development of new secure global agglomerations, particularly of collaborative urban networks between world cities and new mobility systems – biofuels, hydrogen and hybrids – that can guarantee continued interconnection both within and between world cities.

SURI represents an attempt to fix representational space as a space of securitized urban ecologies. In doing so, it seeks to present *the* solutions to the new pressures bearing down on relationships between cities, infrastructures and resource flows. The narrow relational space of its constitution remains largely 'hidden'. World cities are the critical sites for the production of new fixes and strategies designed to attempt to ensure their continued reproduction. Coalitions of interests, that include corporate interests, urban political elites and elements of the environmental movement are actively – if not always aware of the consequences – seeking to produce *the* agenda about the ways in which the maintenance and enhancement of economic growth can be achieved in a context where resource constraints and the challenges of climate change might intuitively be viewed as a threat to economic growth. They are producing an increasingly accepted and dominant view about the primacy of their own role and that of particular world cities in 'effectively' managing the tensions between urban economic growth, territorial expansion, resource constraint and climate change.

Attempts to build UES through the development of strategies of SURI in existing world cities are informing a dominant and largely unchallenged view of the relationship between cities and infrastructure networks. Yet, by questioning the inevitability of these responses we open up not only the constrained capacities to develop responses in relation to ordinary cities but also the potential for very different types of responses that have a much closer understanding of the 'needs' of particular urban contexts. SURI can be seen as representational not only in this respect but also in respect of its relative positioning of its representations vis-à-vis other possibilities within world cities and in non-world cities.

There are significant and fundamental implications of these new styles of response in world cities that are being developed by narrow cadres of social interests. We believe it is imperative that there is a wider critical engagement with the implications of these responses – whom they benefit, how, who is included and excluded – as a much needed counterweight to the overwhelmingly positive and uncritical treatment of these productionist-style responses. We are concerned about this and have sought to develop an appreciation of the political implications of largely unchallenged, depoliticized visions of the future of cities, infrastructures and resource flows. Most significantly, we are concerned with seeking to deepen understanding of the extent to which these responses signify the metropolitanization of ecological resources which is underpinned by the seemingly paradoxical representations of building relative autonomy for world cities while also seeking to build new forms of connection between world cities. In this two new scales become critically important – the degree to which the metropolitan area is more self-reliant and simultaneously the extent to which ecologically secure interurban networks connect world cities – captured in the vision of a secure archipelago of networked ecologies. World cities have become increasingly concerned about the question of systemic preparation for a period of climate change and resource constraint and in assessing the degree to which cities can continue to 'guarantee' their material reproduction. This, as we have argued, is not an inevitable or natural development but is an achievement or an accomplishment that embodies and exemplifies particular coalitions of interests and that is being played out in a narrow class of world cities. In doing so these responses seek to underpin narrower forms of bounded security in relation to these particular cities, social interests, and those like them.

The production of the dominant discourse is predicated on a view that it is cities and particular world cities that are the motors of economic growth in a globalized economy, that the interconnectivity between these cities is crucial to their economic reproduction, and that in an era of resource constraint and political instability around supplies of critical resources strategic responses become crucial. World cities are positioned, and are positioning themselves – through political representatives and collaborative networks – as being key producers and potential victims of climate change and also significant resource consumers. Coalitions of interests are positioning world cities as the spaces and places for the maintenance and enhancement of economic growth and the infrastructural and resource innovations necessary to capture the resources to underpin this growth while also recognizing a necessity to respond to the challenges of climate change.

Rethinking UES: building socially progressive and collective planetary security

New pressures to reconfigure city, infrastructure and resource relationships are becoming manifest at a point in history when the governance of these relationships is increasingly polycentric, at multiple levels or scale of governance and control is dispersed and distributed. It is within this context that the forms of governance that are emerging cannot be left to narrowly constituted social interests but need to be broadly designed and experimented with to 'effectively' intervene. The development of responses that offer an alternative to SURI requires the creation of new relational spaces of possibility. This means the creation of new organizational spaces that bring together social interests that extend and/or differ from those producing SURI. It also means creating a context within which not only a broad range of different social interests can come together but recognizing that these different social interests come with a variety of motivations, expectations, forms of knowledge and finance.

Here we seek to address this critical gap by developing a framework for understanding the participatory possibilities of purposive urban transitions. What is particularly interesting about UES and SURI is that territorial priorities at the scale of the city – economic growth targets, carbon emissions reduction aspirations and so on – are becoming strategically intertwined with the reconfiguration of socio-technical infrastructure systems and these systems may or, more often, may not be organized at the scale of the city. Consequently there needs to be a wider debate about the relationship between cities and their food, energy, water, waste and transport regimes. Our analysis of UES and SURI has led us to speculate about what a manifesto for an alternative form of UES might look like, that sought to develop a more socially progressive and collective planetary security through different types of urban responses.

In doing this we have developed a framework for thinking about the future development of cities in period of global ecological change that is informed by six issues:

1. How ecological pressures are understood.
2. The political-ecology of existing resource flows.
3. Whose social visions it is that informs a reconfigured urbanism.
4. The types of new relational networked ecologies that are envisaged and how they are negotiated.
5. The ways in which capacity and capability are developed to inform socio-technical and configurational change.
6. The forms of societal learning that take place and how they reshape visions and networked ecologies.

First, there is a need for a more relational understanding of global ecological pressures that acknowledges that these pressures will reshape the wider material context within which cities reproduce themselves but that also explicitly acknowledges the ways in which the metalogistical systems that support cities, through multiple material flows, are also responsible for reshaping planetary ecology. Such an understanding needs to prioritize how global ecological change differentially reshapes the lives of different sectors and residents within cities – which identifies the socio-economic groups that are most vulnerable and where is the least capacity to respond – to climate change and the implications for resource flows. It also must address the global and biospherical implications of the city and what the global implications are of the metalogistical systems and resource flows in and out of the city. The challenge is to develop a relational understanding of pressures that acknowledges the dual and ambivalent role of the city, as both a victim and cause of global ecological change, but to do so in a way that directly explores the intra-urban implications of such a change and the ways in which it reshapes global ecological processes.

Second, understanding the political-ecology of existing resource flows refers to the process of trying to make more visible and explicit the social and technical organization of resource flows into and out of cities. It means understanding who controls networks, how they are organized as commodities or public goods, how it is that access to various types of resource flows are organized within the city in terms of differential access to food, energy, water and mobility; externally how these resource flows connect cities to other scales regionally, nationally and internationally and the forms of interdependencies that are created through these flows. Laying bare the societal organization and differential distribution of resource flows makes it possible to begin to understand the degree of lock-in and obduracy in the organization of flows in relation to the city. It also exposes the extent to which the social interests implicated in resource flows are open to urban priorities and in the reshaping of resource flows in systemic ways. In doing this it helps to develop a better understanding of how changed global conditions may reshape the longer-term reliability and resilience of these flows.

Third, in thinking through what an alternative urban transition would look like it is necessary to understand the extent to which there is a common and shared understanding among the wide range of social interests that produce territorial priorities and socio-technical systems of energy, water, waste and transport infrastructures. Visions offer the potential to both constitute and to present a shared and broader understanding of territorial and regime interests. Visions of an urban infrastructure transition may bring together both an understanding of the changes envisaged in an infrastructure system over time but do so in relation to territorial priorities. The production of visions is an important participatory process used to engage, inspire and mobilize a wide variety of different social actors but involves negotiation and struggle. In terms of urban infrastructure this may involve broader coalitions, including representatives of utilities, municipal government, regulators, developers, business, citizens, users and so on. Visions and the goals they outline provide a reference point through which networks can be built, gaining commitments to participate, orientating the actions of potential participants and constituencies, and in persuading potential participants of the desirability of transition. Although visions are not fixed and will change over time with the variety of social interests who become involved, this is currently articulated around narrow coalitions of interests and must be based on a more broadly constituted sense of what a purposive urban transition would look like.

Fourth, there is a need to develop alternative conceptions of resource interdependency and in producing visions to establish negotiated relations with a wider range of social interests 'external' to the city. Whereas SURI is characterized by transcendence of ecological constraints, a productionist bias and the development of internalized resource flows an alternative logic would begin to recognize absolute limits to development and resource consumption. Rather than a production-related focus on new technological solutions an alternative would recognize the critical role of behavioural and cultural change in managing demand downwards. Rather than developing internalized resource flows that withdraw from existing infrastructure the alternative would attempt to develop much more negotiated responses to the external societal and social interests that provide the resource flows for cities and deal with urban wastes. Such a negotiation would seek to develop different ways of thinking about new resource interdependencies that acknowledge limits and responsibilities and develops fair shares and fairer trade.

Reconstructing a vision of an urban infrastructure transition encompasses multilevel governance arrangements and socio-technical systems that are often messy involving multiple actors and institutions across different scales. But, when addressing systemic transitions in urban infrastructures effectively it is necessary to involve representatives of a wide variety of relevant social interests.

Fifth, the development of capacity and capability to manage socio-technical change and experiment needs to be promoted rather than focusing primarily on technological and productionist projects. There has been a tendency in SURI to focus upon decentralized technologies and technologies of resource production and transformation rather than upon wider issues of social, cultural and behavioural change in developing new practices that could keep a downward pressure on resource consumption. This requires the development of 'intermediary' capacity and capability to engage with users, publics and different socio-economic groups in undertaking niche activities and experiments in new styles of urban living. Rather than seeking to inject new technologies into infrastructure systems this would mean developing new social practices, routines and behaviours that would aim to maintain a downward pressure on resource use.

Finally, the development of more reflexive styles of policymaking and societal learning about the implications of socio-technical experimentation of urban infrastructure are required. This is necessary to inform notions of the reshaping of change in the organization of urban resource flows. It becomes critical at the level of cities to learn about how to reshape infrastructure systems and, in particular, to use experiments, demonstrations and niches as contexts for wider societal learning about both the malleability and obduracy of changing socio-technical systems at the urban level. In part this is about understanding what the boundaries and limits are for systemic change within cities – or, in other words, in how far it is possible to make widespread changes in an infrastructure and to what extent other sets of social interests and agencies at other scales need to reconfigure regulations, policies and priorities to enable cities to shape their infrastructure systems. As well as learning about the malleability of urban systems, developing linkages with infrastructure systems at other scales it also means attempting to develop networks between cities that are not exclusive or exclusionary in order to share different types of relevant practices rather than develop a singular fix.

Our most significant concern is how the pressures of UES and SURI as a strategic response are designed to secure the divisible securitization of resources at the metropolitan scale through world cities rather than actively contributing to wider collective solutions to the implications of climate change and resource constraint. While we have sought to offer clarity about our own position we also recognize that we are dealing with a set of social visions and socio-technical aspirations that have to be actively translated into social action. While we firmly believe that there is already sufficient evidence to support our own critical position on the economic motivations and emerging consequences of UES we also recognize that there is a need for ongoing conceptual and empirical work to generate further understanding of these developments. Empirical investigation of the emergence of UES and SURI in several world city contexts, focusing on the resonances and dissonances between pressures and different responses, is necessary. As is further conceptual and theoretical analysis about whether UES and SURI represents a particular political, ecological and socio-technical response to questions of climate change and resource constraint under conditions of neoliberal urbanism. Finally, wider research and political engagement is required to highlight the visibility and potential of resistance to UES and alternative strategies underpinned by commitments to mutual interdependencies, fair shares and social needs. These, we argue, are the critical issues raised by urban ecological security.

Glossary

Alternative logics refers to the development of 'alternative' urban responses to climate change and resource constraint in contrast to the logic of SURI. These usually exhibit one or more of the following characteristics: a wider and more inclusive social vision for the development of urban resources not solely based on an economic or security logic; the involvement of a much wider group of stakeholders in the development of the process of formulating visions and the capacity to act – usually involving a wider range of publics, communities and stakeholders; a greater emphasis on the development of social, cultural and behavioural change rather than a focus on technocratic responses; where technologies are involved there is usually a greater focus on issues of social and local control. 'Alternatives' usually take a broader interest in questions and dialogue around collective planetary security, fair shares and mutual interdependencies.

Anthropocene is the latest geological period that can be characterized by a series of physical markers that are caused by humans. The Greek origins of the word are 'anthropo-' meaning 'human' and '-cene' meaning 'new.' It was officially proposed in 2008 as a new geological era that was considered to have started in the late eighteenth century when the activities of humans first began to have a significant global impact on the Earth's climate and ecosystems.

Archipelagos of global infrastructure are the globally oriented communications infrastructures of telecommunications, international airlines, freight logistics and international shipping systems that connect together and are located within or adjacent to world cities. These systems can be seen as connecting islands of world cities together in an interconnected global infrastructure system.

Autonomous cities, closed cities, relocalization, self-reliant cities refers to attempts to cease – or more usually radically reduce – a city's reliance on the external importation (and exportation of wastes and resources) of critical resources flows, including energy, water, food, wastes and so on. This usually involves a combination of strategies – demand-side management, the development of local resources, production and reuse technologies, and withdrawal from reliance on external infrastructures. Such strategies are often called the development of circular urban metabolisms. Achieving a totally closed or autonomous system is virtually impossible due to leakages but a city may be able to achieve greater relative self-reliance.

Bypassing and decoupling from national infrastructure underlies attempts by cities seeking to develop more autonomous or self-reliant infrastructure by withdrawal from reliance on existing or incumbent infrastructure networks (at different scales regional, national and international) that currently supply the city. This can be done in a number of different ways: by aiming to develop internal supplies within the city (i.e. local generation and desalination, demand management) rather than investing in infrastructure or supply technologies external to the city; through seeking to drive infrastructure investment into metropolitan areas (instead of external areas) to strengthen the capacity of the city grid to develop new supply; to work to develop infrastructure resources close to the city to increase the security of the metropolitan network (e.g. develop on city-owned land, support private sector investors etc). There are of course issues of resistance, tensions and conflicts about the politics of infrastructure withdrawal and bypass.

Cities as planetary terraciders refers to the role of cities in changing global ecological conditions through overconsumption of resources and production of pollutants and carbon emissions that potentially leads to the destruction of global ecologies.

Cities as planetary terraformers addresses the role of cities in creating 'environments' for an increasingly large proportion of the world's population, estimated at 50 per cent in 2005. Cities have in this sense been remarkably successful in building massive urban agglomerations that are able to, temporarily, support millions of people. See, for example, Tokyo's population of 30 million.

Collective security refers to attempts to build shared or collective security at a global scale. Rather than bounding security at a particular scale, this looks seriously at existing differences in access to critical resources, differences in the vulnerability and resilience of places and populations, and the uneven interdependencies between places through resource flows and, in doing so, at attempts to build security through the creation of fair shares, mutual interdependencies and some recognition of limits and justice in allocation of resources.

Competing transitions are the coexisting and competing transition pathways for urban infrastructure. For example, the aspiration of a low carbon transition could be achieved in multiple ways – either through systemic changes in centralized networks or alternatively through an expansion of decentralized energy technologies within cities. Consequently there may be tensions and conflicts between urban aspirations and national transitions.

Divisible or bounded security refers to attempts to construct security at a particular spatial scale, whether that be individual buildings, neighbourhoods, cities, regions or nations. A particularly useful example would be ecological extensions of gated communities that layer on local energy, water, food production and waste disposal in order to attempt to guarantee security and longer-term reproduction. Such efforts should be regarded as aspirations and attempts to guarantee security as they inevitably have perverse and contradictory effects.

Eco-competitive urbanism encompasses the increasing attempts to reconnect urban responses to the global economic crisis with responses to the ecological crisis. This manifests itself in a number of different ways, most notably: the green jobs movement; attempt to develop low carbon economies; and incorporating ecological dimensions into indexes that compare the economic performance of cities.

Ecological enclaves are the creation of bounded enclaves that are able, albeit on a temporary basis, to provide a degree of ecological closure, through circular metabolisms, local food production systems, protection from climate change and so on. These are the spaces created through rebundled network ecologies and integrated urbanism.

Global urban agglomerations are the parallel development of global infrastructure networks that connect world cities through metalogistical systems. They also refer to the networks of world cities that are attempting to develop emblematic and exemplary infrastructure solutions.

Glurbanization of networked ecologies is the parallel rescaling of urban network ecologies through the development of (1) more self-reliant world cities that are also (2) better connected with other world cities through globally orientated infrastructures.

Integrated eco-urbanism refers to urban development projects at a range of different scales (houses, villages, towns and cities) that seek to integrate demand management and local production technologies within the architectural and engineering design of new-build projects to develop more resilient and self-reliant styles of urbanism. Different styles of integrated eco-urbanism include eco-towns, Eco-Blocks, eco-islands, eco-cities and eco-regions.

Interconnected networks are the tightly coupled interconnections between infrastructure networks. One example of this is the critical role of information technology and energy systems in the effective operation of all other infrastructure networks. Disruptions and/or breakdowns in these core infrastructures can lead to cascading failures in other infrastructure networks.

Limits of bounded autonomy and new interdependencies refers to the political and physical limits of creating tightly bounded or closed systems. Evidence shows that it is extremely difficult, if not impossible, to create autonomous fixes for infrastructure. Even eco-cities are assumed to have high levels of connectivity, within and between cities, even if they are able to build a greater degree of self-reliance in their infrastructures systems. Instead more bounded cities create new interdependencies rather than total autonomy.

Metalogistical systems are the global systems and flows of energy, food, materials, water and wastes resources, people, goods and services that are heavily focused on cities as environmental transformers of resources that have global ecological implications. World cities have a critical role as key nodes in metalogistical systems in both a physical sense and through critical control functions of global systems (Luke 2003).

Metropolitanization of networked ecologies are the processes through which world cities seek to subject networked ecologies to greater scrutiny to understanding existing flows; to understand how such ecologies are likely to change over time; and to influence their reconfiguration – i.e. through decentralization; to develop resources close to but not within cities; to maintain connectivity between world cities; and finally to develop new networked technologies in collaboration with other world cities.

Networked ecologies refers to the increasing visibility of the ecological implications of the infrastructure systems that constitute the networked society. In particular what becomes visible are the material and social interdependencies created by existing production and consumption systems; the global ecological consequences – especially carbon emissions – associated with flows; and, the vulnerability of resource flows in the longer-term.

New urban network technologies can be understood as the efforts of cities to develop alternative energy networks – hydrogen, biofuels and district heating and cooling schemes – grey water and recycled water schemes, that are retrofitted into urban contexts partly to develop new infrastructure systems.

Projects, demonstrations, niches and experiments are the development of smaller scale projects that are designed to act as demonstrations, niches or experiments of responses to climate change and resource constraint. These exemplify smaller scale examples of systemic changes that could be rolled out over the whole urban system.

Rebundling of networked ecologies refers to the process of the selective repackaging of infrastructure networks, material resources and the built environment at particular scales to produce rebundled network ecologies. For example, local food production systems, energy autonomous buildings, water and waste recycling technologies and so on.

Reconfiguration of existing infrastructure are attempts to reshape particular networked technologies in a specific way and are primarily focused on water, food, and energy and flood security. Here the conception is on preparing an urban infrastructure network to deal with the implications of changes, ecological and economic conditions, through systemic attempts to reconfigure the resource flows, technologies and aspects of the social organization of network infrastructures.

Resource constraint refers to the ways in which climate change and carbon regulation are layered over existing political and economic systems and may reshape the longer-term security, availability and reliability of critical resource flows through cities including energy, water, food, and waste disposal.

Secure urbanism and resilient infrastructure refers to the response in world cities to climate change and resource constraint involving three sets of strategies: strategic protection: the development of self-reliance; and the building of global urban agglomerations.

Strategic 'protection' of cities is the process of developing knowledge and expertise to understand the longer-term implications of climate change and resource constraint on world cities. In particular this includes the local repercussions of climate change; the regional implications for water, waste and food; the implications of carbon regulation, and the wider ramifications for the availability and reliability of energy, mobility and food systems. This marks an extension of conventional military and security concerns into the resilience of resource flows.

Systemic urban transitions refers to systemic change in an existing city through the use of new concepts such as low carbon cities, zero carbon cities, peak oil cities and so on. These are simultaneously new societal visions of contemporary urbanism and also set up new technological expectations about how cities, buildings, infrastructures, institutions, publics and stakeholders and so on might be involved in a wider system change in response to new ecological pressures.

Transcendent urbanism the claim is that new styles of integrated eco-urbanism are, through water, energy, waste and carbon neutrality, able to transcend and overcome what would be considered to be conventional ecological limits and/or constraint to normal styles of urban development.

Uneven capacity and capability is the uneven development of knowledge, expertise, resources, personnel and social institutions able to develop systemic and/or project-based capacity to develop responses to climate change and resource constraint.

Urban adaptation is the adjustment in urban socio-material systems in response to climate pressures in ways that moderate harm or exploits beneficial opportunities. Various types of adaptation can be distinguished, including anticipatory and reactive adaptation, private and public adaptation, and autonomous and planned adaptation.

Urban adaptive capacity is the capability of a city to adjust to climate change (including climate variability and extremes), to moderate potential damages, to take advantage of opportunities, or to cope with the consequences.

Urban carbon regulation refers to national state roles in developing subnational carbon management or reduction targets that are passed on to cities and city-regions for compliance.

Urban climate impacts are the consequences of climate change on natural and human systems in cities. Potential impacts are impacts that may occur given a projected change in climate, without considering adaptation. Impacts after adaptation are called residual impacts.

Urban ecological 'fixes' refers to the selective incorporation of ecological goals in the greening of urban governance. It draws attention to the particular dilemmas urban regimes in different cities currently face in balancing economic, social and environmental demands (see While et al. 2004).

Urban ecological security refers to the selective incorporation of ecological issues into wider strategic debates about the approaches and responses required by cities in order to attempt to ensure their continued social, economic and material reproduction of cities in the context of global ecological change.

Urban intermediaries are the institutions that seek to develop capacity and capability to develop transitions in infrastructure networks. These can be either at the level of standalone projects and demonstrations in a network or the development of projects that can be upscaled to establish systemic change.

Urban reproduction refers to the social, economic and ecological processes that enable the effective reproduction of cities and to meet 'their' wider growth objectives and territorial priorities.

Urban resilience is normally understood as the amount of change a city can undergo without changing state. This could include (1) the capacity of a system to tolerate disturbance without changing state, and (2) the ability to recover from the effect.

Urban retrofitting refers to strategies and initiatives for the city to become more sustainable, through new technologies and changed social practices, by redesigning existing infrastructure and the built environment in a city.

Urban sensitivity is the degree to which a city is affected, either adversely or beneficially, by climate-related stimuli. The effect may be direct (changing demand for water or variability of temperature) or indirect (such as damages caused by an increase in the frequency of coastal flooding due to sea level rise).

Urban transitions refers to the development of purposive and managed socio-technical change in an urban infrastructure network. This is usually characterized by the development of a wider urban visions (e.g. low carbon transition); the enrolment of multiple stakeholders and publics;

and the development of intermediary capability to manage systemic change in the socio-technical organization of a network. A successful transition would be characterized by the above and material changes in the organization of an infrastructure system through, for example, reduced energy consumption, carbon intensity and so on.

Urban vulnerability is the degree to which a city is susceptible to, or unable to cope with, adverse effects of climate change, including climate variability and extremes.

Urbanatura refers to the increasingly variable and uncertain global ecological conditions of the anthropocene within which 'cities' must attempt to ensure their continued social, economic and material reproduction (see Luke 2008).

World cities are the cities that function as important nodes in the global economic system. World cities have a direct and tangible effect on global affairs through the control of social networks, economic relationships and material infrastructures.

Notes

Chapter 2

1. Global Security Newswire, 20 May 2008.
2. See www.c40cities.org (accessed 30 March 2010).

Chapter 3

1. See http://bie.berkeley.edu/ecoblocks (accessed 21 September 2009).
2. See www.telegraph.co.uk/news/worldnews/australiaandthepacific/tuvalu/5871093/Tuvalu-plots-worlds-first-zero-carbon-output-by-2020.html (accessed 25 November 2009).
3. See www.citymayors.com/environment/biofuels-usa.html (accessed 30 September 2009).
4. See www.biofuel-cities.eu/index.php?id=3252 (accessed 30 September 2009).
5. See www.fuel-cell-bus-club.com/index.php?module=pagesetter&func=viewpub&tid=1&pid=11 (accessed 16 August 2007).
6. See www.hydrogenhighway.ca.gov/vision/vision.pdf (accessed 16 August 2007).
7. See www.clw.csiro.au/division/adelaide/seminars/StormwaterSeminar20071007.pdf (accessed 28 September 2009).
8. See www.sydneyolympicpark.com.au/education_and_learning/environment/water/wrams (accessed 28 September 2009).
9. See http://news.bbc.co.uk/1/hi/england/8246009.stm (accessed 28 September 2009).
10. See www.thisislondon.co.uk/standard-mayor/article-23694458-boris-johnson-uk-will-lead-world-on-electric-cars.do;jsessionid=1EB4C78BC7E3AB656A8DD1E50DE364F4 (accessed 30 March 2010).
11. See www.washingtonpost.com/wp-dyn/content/article/2007/03/23/AR2007032301753.html (accessed 28 September 2009).
12. See www.cpfund.ca (accessed 28 September 2009).

13. See www.hearthmakers.org (accessed 28 September 2009).
14. See http://ngea.org (accessed 28 September 2009).
15. See www.energysavingcommunity.co.uk/community-energy-networks.html (accessed 28 September 2009).
16. See www.wwfchina.org/english/sub_loca.php?loca=1&sub=96 (accessed 25 September 2009).
17. See www.wwfchina.org/english/sub_loca.php?loca=1&sub=96 (accessed 25 September 2009).
18. See www.portlandpeakoil.org/content/reduce-total-oil-and-natural-gas-consumption-50-percent-over-next-25-years (accessed 12 January 2009).
19. See www.environment.gov.au/water/programs/urban/pubs/desalination-guidelines.pdf (accessed 25 September 2009).
20. See www.urbanwateralliance.org.au (accessed 25 September 2009).
21. See http://meetingorganizer.copernicus.org/EGU2009/EGU2009-7473-1.pdf (accessed 28 September 2009).
22. See www.unesco-ihe.org/About/News/Inaugural-address-of-Prof.-Chris-Zevenbergen-14-December (accessed 28 September 2009).
23. See www.european-waternews.com/news/id611-Road_Map_Towards_a_Flood_Resilient_Urban_Environment.html (accessed 28 September 2009).
24. See www.local.org/independ.html (accessed 25 November 2009).
25. See, for example, www.foodsecurity.org/PrimerCFSCUAC.pdf (accessed 28 September 2009).
26. See www.foodsecurity.org/PrimerCFSCUAC.pdf (accessed 28 September 2009).

Chapter 4

1. See www.c40cities.org (accessed 28 September 2009).
2. See www.thamesweb.com (accessed 28 September 2009).
3. See www.kensetsu.metro.tokyo.jp/c40/act6_E/practice01.html (accessed 8 September 2009).
4. See www2.kankyo.metro.tokyo.jp/c40/c40tokyo/pdf/jointactionsummary.pdf (accessed 8 September 2009).
5. See www.seattle.gov/council/issues/zerowaste.htm (accessed 8 September 2009).
6. See www.c40cities.org/bestpractices/waste/sydney_recycling.jsp (accessed 8 September 2009).
7. See www.sustinablecity.org/plansolidwst/intro.htm (accessed 8 September 2009).
8. See www.c40cities.org (accessed 30 March 2010).
9. The Climate Group, www.theclimategroup.org/index.php?pid=734 (accessed 8 December 2006).
10. The cities were San Francisco, Seattle, Mexico City, Chicago, Toronto, New York City, London, Barcelona, Paris, Copenhagen, Berlin, Cape Town, Beijing, Tokyo and Melbourne.
11. See www.theclimategroup.org/assets/Cities_publication.pdf (accessed 13 December 2006).
12. From authors' programme of interviews with urban and national policymakers.

13. See www.iclei.org (accessed 8 September 2009).
14. See www.theclimategroup.org/assets/July_31_Overview_.pdf (accessed 13 December 2006).
15. See www.theclimategroup.org/index.php?pid=838 (accessed 13 December 2006).

Chapter 5

1. See http://relocalize.net/about/relocalization (accessed 29 January 2008).

References

Abu-Lughod, J. (1991) Going Beyond Global Babble. In King, A.D. (ed.) *Culture, Globalization and the World System*. Minneapolis, MN: University of Minnesota Press, pp. 131–138.

Anderson, B. (1991) *Imagined Communities: Reflections on the Origin and Spread of Nationalism*. London: Verso.

Arup (2005) *Arup Unveils Plans for World's First Sustainable City in Dongtan, China*. Available from: www.arup.com/arup/newsitem.cfm?pageid=7009 (accessed 6 May 2008).

Arup (2006) *London Looks to the East for Inspiration to Cut Emissions*. Available from: www.arup.com/newsitem.cfm?pageid=8295 (accessed 6 May 2008).

Barry, A. and Slater, D. (2002) Introduction: The Technological Economy. *Economy and Society* 31(2): 175–193.

Beaverstock, J.V., Smith, R.G. and Taylor, P. J. (1999) A Roster of World Cities. *Cities* 16(6): 445–458.

Bicknell, J., Dodman, D. and Satterthwaite, D. (2009) *Adapting Cities to Climate Change: Understanding and Addressing the Development Challenges*. London: Earthscan.

Bijker, W.E., Hughes, T.P. and Pinch, T. (eds) (1987) *The Social Construction of Technological Systems*. Cambridge, MA: MIT Press.

Bohman, J. (1999) Practical Reason and Cultural Constraint. In Shusterman, R. (ed.) *Bourdieu: A Critical Reader*. Oxford: Blackwell.

Bourdieu, P. (1986) The Forms of Capital. In Richardson, J.G. (ed.) *Handbook for the Theory and Research for the Sociology of Education*. New York: Greenwood.

Bourdieu, P. (1991) *Language and Symbolic Power*. Edited and introduced by Thompson, J.B. Trans. Raymond, G. and Adamson, M. Cambridge: Polity.

Bourdieu, P. (1993) *Sociology in Question*. London: Sage.

Bourdieu, P. (2000) *Pascalian Meditations*.Translated by Nice, R. Cambridge: Polity.

Brand, P. (2007) Green Subjection: The Politics of Neoliberal Urban Environmental Management. *International Journal of Urban and Regional Research* 31(3): 616–632.

Brenner, N. (2004) *New State Spaces: Urban Governance and the Rescaling of Statehood*. New York: Oxford University Press.

Brenner, N. and Keil, R. (eds) (2006) *The Global Cities Reader*. London: Routledge.

Brenner, N. and Theodore, N. (2002) Preface: From the 'New Localism' to the Spaces of Neoliberalism. *Antipode* 34(3): 341–347.

Bulkeley, H. (2005) Reconfiguring Environmental Governance: Towards a Politics of Scales and Networks. *Political Geography* 24(8): 875–902.

Bulkeley, H. and Betsill, M. (2003) *Cities and Climate Change: Urban Sustainability and Global Environmental Governance*. London: Routledge.

Burton, M.L. and Hicks, M.J. (2005) *Hurricane Katrina: Preliminary Estimates of Commercial and Public Sector Damages*. Huntington, VA: Center for Business and Economic Research, Marshall University. Available at: www.marshall.edu/cber/research/katrina/Katrina-Estimates.pdf (accessed 30 March 2010).

C20 (2005) *Low Carbon Leader: Cities Oct. 2005*. London: Climate Group.

California Environmental Protection Agency (2005) *California Hydrogen Blueprint Plan*, Volume 1, May. Available at: www.hydrogenhighway.ca.gov/plan/reports/volume1_050505.pdf (accessed 30 March 2010).

Callon, M. (1998a) Introduction: The Embeddedness of Economic Markets in Economics. In Callon, M. (ed.) *The Laws of the Markets*. Oxford: Blackwell, pp. 1–57.

Callon, M. (1998b) An Essay on Framing and Overflowing: Economic Externalities Revisited by Sociology. In Callon, M. (ed.) *The Laws of the Markets*. Oxford: Blackwell, pp. 244–269.

Cameron, A. and Palan, R. (2004) *The Imagined Economies of Globalization*. London: Sage.

Castells, M. (2000) *The Information Age: Economy, Society and Culture, Volume One: The Rise of the Network Society*, 2nd edn. Oxford: Blackwell.

Cisneros, B.J. and Rose, J.B. (eds) (2005) *Urban Water Security: Managing Risks*. UNESCO-Ihp. New York: Routledge.

City of Melbourne (n.d.) *Climate Neutral Water Saving Schemes*. Available at: www.melbourne.vic.gov.au/Environment/WhatCouncilisDoing/Documents/Climate_neutral_water_saving_schemes.pdf (accessed 30 March 2010).

Coaffee, J. (2008) Risk, Resilience, and Environmentally Sustainable Cities. *Energy Policy* 36(12): 4633–4638.

Dalby, S. (2007) Anthropocene Geopolitics: Globalisation, Empire, Environment and Critique. *Geography Compass* 1(1): 103–118.

Department for Communities and Local Government (DCLG) (2009) *Planning Policy Statement: Eco-towns*. London: DCLG, July.

Department of Energy and Climate Change (DECC) (2009) *The UK Low Carbon Transition Plan National Strategy for Climate Change*. London: DECC, July.

Department of Homeland Security (DHS) (2003) *Homeland Security Presidential Directive 7: Critical Infrastructure Identification, Prioritization, and Protection*. Available from: http://www.dhs.gov/xabout/laws/gc_1214597989952.shtm (accessed 27 April 2010).

Department of Sustainability and Environment (2002) *Melbourne 2030: Planning for Sustainable Growth*. Available from: www.dse.vic.gov.au/melbourne2030online/content/site_functions/pdfs.html (accessed 30 March 2010).

Desfor, G. and Keil, R. (2004) *Nature and the City: Making Environmental Policy in Toronto and Los Angeles*. Tuscon, AZ: University of Arizona Press.

Dodman, D. (2009) Blaming Cities for Climate Change? An Analysis of Urban Greenhouse Gas Emissions. *Environment and Urbanization* 21(2): 545–567.

Environment Agency (2007) *Towards Water Neutrality in the Thames Gateway*. Science Report SC060100/SR3. Available at: http://publications.environment-agency.gov.uk/pdf/SCHO1107BNMN-e-e.pdf?lang=_e (accessed 30 March 2010).

Fischer, F. (2003) *Reframing Public Policy: Discursive Politics and Deliberative Practices*. New York: Oxford University Press.

Foucault, M. (2008) *Security, Territory, Population*. Basingstoke: Palgrave-Macmillan.

Fraker, H. (2006) Unforbidden Cities: Can a New Type of 'Gated Community' Reverse China's Ecological Debacle? *California Magazine* 118(5): 44–49.

Friedmann, J. (1986) The World City Hypothesis. *Development and Change* 17(1): 69–83.

Geels, F.W. and Schot, J. (2007) Typology of Socio-technical Transition Pathways. *Research Policy* 36(3): 399–417.

Goffman, E. (1974) *Frame Analysis: An Essay on the Organization of Experience*. New York: Harper & Row.

Graham, S. (2004) Cities as Strategic Sites: Place Annihilation and Urban Geopolitics. In Graham, S., *Cities, War and Terrorism*. Oxford: Blackwell, pp. 31–53.
Graham, S. (2006) Cities and the 'War on Terror'. *International Journal of Urban and Regional Research* 30(2): 255–276.
Graham, S. and Marvin, S. (2001) *Splintering Urbanism: Networked Infrastructures, Technological Mobilities and the Urban Condition*. London: Routledge.
Harding, A. (1995) Elite Theory and Growth Machines. In Judge, D., Stoker, G. and Wolman, H. (eds) *Theories of Urban Politics*. London: Sage, pp. 35–53.
Hardoy, J.E., Mitlin, D. and Satterthwaite, D. (2001) *Environmental Problems in an Urbanizing World*. London: Earthscan.
Harvey, D. (1989) From Managerialism to Entrepreneurialism: The Transformation in Urban Governance in Late Capitalism. *Geografiska Annaler* Series B, 71(1): 3–17.
Harvey, D. (1990) *The Condition of Postmodernity*. Oxford: Blackwell.
Harvey, D. (2006) *Spaces of Global Capitalism: Towards a Theory of Uneven Geographical Development*. London: Verso.
Heynen, N., Kaika, M. and Swyngedouw, E. (2006) *In the Nature of Cities: Urban Political Ecology and the Politics of Urban Metabolism*. London: Routledge.
Hodson, M. and Marvin, S. (2007) Transforming London/Testing London: Understanding the Role of the National Exemplar in Constructing 'Strategic Glurbanisation'. *International Journal of Urban and Regional Research* 31(2): 303–325.
Hodson, M. and Marvin, S. (2009a) 'Urban Ecological Security' A new urban paradigm? *International Journal of Urban and Regional Research* 33(1): 193–215.
Hodson, M. and Marvin, S. (2009b) Cities Mediating Technological Transitions: Understanding Visions, Intermediation and Consequences. *Technology Analysis and Strategic Management* 21(4): 515–534.
Hodson, M. and Marvin, S. (2010) Can cities shape socio-technical transitions and how would we know if they were? *Research Policy* 39(4): 477–485.
Hopkins, R. (2008) *The Transition Handbook: From Oil Dependency to Local Resilience*. Totnes: Green Books.
Hughes, T.P. (1987) The Evolution of Large Technological Systems. In Bijker, W.E., Hughes, T.P. and Pinch, T.J. (eds) *The Social Construction of Technological Systems: New Directions in the Sociology and History of Technology*. Cambridge, MA: MIT Press, pp. 17–50.
Intergovernmental Panel on Climate Change (IPCC) (2001) *Third Assessment Report: Climate Change 2001 (TAR)*. Cambridge: Cambridge University Press for IPCC.
Intergovernmental Panel on Climate Change (IPCC) (2007) *Fourth Assessment Report: Climate Change 2007 (AR4)*. Cambridge: Cambridge University Press for IPCC.
Jessop, B. (2001) Institutional Re (turns) and the Strategic-relational Approach. *Environment and Planning A* 33(7): 1213–1235.
Jessop, B. (2002) *The Future of the Capitalist State*. Cambridge: Polity.
Jessop, B. (2007) *State Power*. Cambridge: Polity.
Jin, M. and Shepherd, J.M. (2005) Inclusion of Urban Landscape in a Climate Model: How Can Satellite Data Help? *Bulletin of the American Meteorological Society* 86(5): 681–689.
Keil, R. (1995) The Environmental Problematic in World Cities. In Knox, P. and Taylor, P.J. (eds) *World Cities in a World-system*. Cambridge: Cambridge University Press.
Keil, R. (2003) Urban Political Ecology. *Urban Geography* 24(8): 723–738.
King, A. (1990) *Global Cities: Post-Imperialism and the Internationalization of London*. London: Routledge.
Koc, M., MacRae, R., Mougeot, L.J.A. and Welsh, J. (1999) *For Hunger-Proof Cities: Sustainable Urban Food Systems*. Ottawa, ON: International Development Research Centre.
Law, J. (1992) *Notes on the Theory of the Actor Network: Ordering, Strategy and Heterogeneity*. Available from: www.lancs.ac.uk/fass/sociology/papers/law-notes-on-ant.pdf (accessed 30 March 2010).
Lerch, D. (2007) *Post Carbon Cities: Planning for Energy and Climate Uncertainty*. Santa Rosa, CA: Post Carbon Institute.

Little, R.G. (2002) Controlling Cascading Failure: Understanding the Vulnerabilities of Interconnected Infrastructures. *Journal of Urban Technology* 9(1): 109–124.

Local Power (2007) *San Francisco Prepares to Vote on Major 'Energy Independence Project'*. Available from: www.local.org/independ.html (accessed 30 March 2010).

Logan, J. and Molotch, H. (1987) *Urban Fortunes: The Political Economy of Place*. Berkeley, CA: University of California Press.

London Assembly (2005) *London Under Threat? Flooding Risk in the Thames Gateway*. London: Environment Committee, Greater London Authority.

Luke, T. (2003) Codes, Collectivities, and Commodities: Rethinking Global Cities as Megalogistical Spaces. In Krause, L. and Petro, P. (eds) *Global Cities: Cinema, Architecture, and Urbanism in a Digital Age*. Piscataway, NJ: Rutgers University Press.

Luke, T. (2006) 'Global Cities' vs. 'global cities': Rethinking Contemporary Urbanism as Public Ecology. In Brenner, N. and Keil, R. (eds) *The Global Cities Reader*. London: Routledge.

Luke, T. (2008) Climatologies as Social Critique: The Social Construction/Creation of Global Warming, Global Dimming, and Global Cooling. In Vanderheiden, S. (ed.) *Political Theory and Global Climate Change*. Cambridge, MA: MIT Press.

McGranahan, G., Balk, D. and Anderson, B. (2007) The Rising Tide: Assessing the Risks of Climate Change and Human Settlements in Low Elevation Coastal Zones. *Environment and Urbanization* 19(1): 17–37.

Massey, D. (2005) *For Space*. London: Sage.

Massey, D. (2007) *World City*. Cambridge: Polity.

Mayor of London (2006a) *The London Plan, Spatial Development Strategy for Greater London, Greater London Authority* (with early alterations). London: Greater London Authority.

Mayor of London (2006b) *Powering London, into the 21st Century*, March. Available from: www.greenpeace.org.uk/MultimediaFiles/Live/FullReport/7474.pdf (accessed 30 March 2010).

Mayor of London (2006c) *The Mayor of London's Submission to the Energy Review*, April. London: Greater London Authority.

Mayor of London (2007a) *Water Matters: The Mayor's Draft Water Strategy. Draft for Consultation with the London Assembly and Functional Bodies*, March. Available from: www.waterwise.org.uk/images/site/Policy/water%20matters.pdf (accessed 30 March 2010).

Mayor of London (2007b) *Action Today to Protect Tomorrow: The Mayor's Climate Change Action Plan*. London: Greater London Authority.

Mayor of London (2007c) *Making Waste Work in London: The Mayor's Draft. Business Waste Management Strategy*. London: Greater London Authority.

Mayor of London (2008) *The London Climate Change Adaptation Strategy*. London: Greater London Authority.

Meadows, D.H., Meadows, D.L., Randers, J. and Behrens, W.W. (1972) *Limits to Growth*. New York: Potomac Associates.

Nicholls, R.J. (2004) Coastal Flooding and Wetland Loss in the 21st Century: Changes under the SRES Climate and Socio-economic Scenarios. *Global Environmental Change* 14: 69–86.

Olds, K. and Yeung, H. (2004) Pathways to Global City Formation: A View from the Developmental City-State of Singapore. *Review of International Political Economy* 11(3): 489–521.

Organization for Economic Cooperation and Development (OECD) (2007) *Literature Review on Climate Change Impacts on Urban City Centres – Initial Findings, by Alistair Hunt and Paul Watkiss*. Paris: Environment Directorate, OECD.

Pickerill, J. and Maxey, L. (eds) (2009) *Low Impact Development: The Future in our Hands*. Available at: http://lowimpactdevelopment.wordpress.com (accessed 30 March 2010).

Pirages, D. and Cousins, K. (eds) (2005) *From Resource Scarcity to Ecological Security*. Cambridge, MA: MIT Press.

Reeves, H. (2008) *Terracide*. Toronto, ON: Cormorant.

Roberts, P. (2004) *The End of Oil*. New York: Houghton Mifflin.

Robinson, J. (2006) *Ordinary Cities: Between Modernity of Development*. London: Routledge.

Sassen, S. (2001) *The Global City*, 2nd edn. Princeton, NJ: Princeton University Press.

Satterthwaite, D. (2008) Cities' Contribution to Global Warming: Notes on the Allocation of Greenhouse Gas Emissions. *Environment and Urbanization* 20(2): 539–549.

Serre, D. and Barroca, B. (2009) Urban Flood Risk Mitigation: From Vulnerability Assessment to Resilient City. *Geophysical Research Abstracts* 11.

Smith, M.R. and Marx, L. (1994) *Does Technology Drive History? The Dilemma of Technological Determinism*. Cambridge, MA: MIT Press.

Soubbotina, T. (2004) *Beyond Economic Growth: An Introduction to Sustainable Development*. Washington, DC: World Bank.

Stern, N. (2006) *The Stern Review Report on the Economics of Climate Change*, October. Available from: www.hm-treasury.gov.uk/sternreview_index.htm (accessed 30 March 2010).

Swyngedouw, E. (2006) Metabolic Urbanization: The Making of Cyborg Cities. In Heynen, N., Kaika, M. and Swyngedouw, E. (2006) *In the Nature of Cities: Urban Political Ecology and the Politics of Urban Metabolism*. London: Routledge.

Swyngedouw, E. and Heynen, N. (2003) Urban Political Ecology, Justice and the Politics of Scale. *Antipode* 35(5): 898–918.

Thames Estuary 2100 (2007) *Thames Estuary 2100: Phase 2 Report, Executive Summary*. Draft v.2, 14 June. Available from: http://mayor.london.gov.uk/london-plan-eip/docs/briefings/bn14-thames-estuary.pdf.

Thompson, J.B. (1991) Editor's Introduction. In Bourdieu, P., *Language and Symbolic Power*. Edited and introduced by Thompson, J.B. Trans. Raymond, G. and Adamson, M. Cambridge: Polity.

Thompson, J.B. (1995) *The Media and Modernity: A Social Theory of the Media*. Cambridge: Polity.

Tokyo Metropolitan Government (2007) *Tokyo Climate Change Strategy*. Available at: www.kankyo.metro.tokyo.jp/kouhou/english/pdf/TOKYO%20Climate%20Change%20Strategy%202007.6.1.pdf (accessed 30 March 2010).

Tyndall Centre (2008) *Engineering Cities: How Can Cities Grow While Reducing Vulnerability and Emissions? Summary of the Urban Integrated Assessment Facility*. Manchester: Tyndall Centre for Climate Change.

United Nations Environment Programme (UNEP) (2007) *Global Environment Outlook: Environment for Development (GEO-4) Assessment*. Available from: www.unep.org/geo/geo4/media/ (accessed 30 March 2010).

United Nations Population Fund (UNFPA) (2007) *State of World Population 2007: Unleashing the Potential of Urban Growth*. New York: UNFPA.

Victorian Government (2007) *Our Water, Our Future: The Next Stage of the Government's Water Plan*. Melbourne: Department of Sustainability and Environment. Available at: www.ourwater.vic.gov.au (accessed 30 March 2010).

While, A. (2008) Climate Change and Planning: Carbon Control and Spatial Regulation. Viewpoint. *Town Planning Review* 78(6): vii–xiii.

While, A., Jonas, A. and Gibbs, D. (2004) The Environment and the Entrepreneurial City: Searching for the Urban 'Sustainability Fix' in Manchester and Leeds. *International Journal of Urban and Regional Research* 28(3): 549–569.

White, I. and Howe, J. (2005) Unpacking Barriers to Sustainable Urban Drainage Use. *Journal of Environmental Policy and Planning* 7(1): 27-43.

Zalasiewicz, J., Williams, M., Smith, A. et al. (2008) Are We Now Living in the Anthropocene? *GSA Today* 18(2): 4–8.

Index